THE ETHICS OF CULTURAL COMPETENCE IN HIGHER EDUCATION

THE ETHICS OF CULTURAL COMPETENCE IN HIGHER EDUCATION

Edited by

Beverly A. Burnell, PhD, and
Heidi L. Schnackenberg, PhD

APPLE ACADEMIC PRESS

Apple Academic Press Inc.
3333 Mistwell Crescent
Oakville, ON L6L 0A2
Canada

Apple Academic Press Inc.
9 Spinnaker Way
Waretown, NJ 08758
USA

First issued in paperback 2021

Exclusive worldwide distribution by CRC Press, a member of Taylor & Francis Group

ISBN 13: 978-1-77463-285-7 (pbk)
ISBN 13: 978-1-926895-50-5 (hbk)

Library of Congress Control Number: 2014045334

Library and Archives Canada Cataloguing in Publication

The ethics of cultural competence in higher education / edited by Beverly A. Burnell, PhD, and Heidi L. Schnackenberg, PhD.

Includes bibliographical references and index.
ISBN 978-1-926895-50-5 (bound)
1. Education, Higher--Moral and ethical aspects. 2. Teaching--Moral and ethical aspects.
3. Cultural competence--Moral and ethical aspects.
I. Burnell, Beverly A., author, editor II. Schnackenberg, Heidi L., author, editor

LB2324.E85 2014 378.01 C2014-907730-0

Apple Academic Press also publishes its books in a variety of electronic formats. Some content that appears in print may not be available in electronic format. For information about Apple Academic Press products, visit our website at **www.appleacademicpress.com** and the CRC Press website at **www.crcpress.com**

ABOUT THE EDITORS

Beverly A. Burnell, PhD

Bev Burnell is an Associate Professor in Counselor Education at SUNY Plattsburgh, New York, and also serves as the Employee Assistance Coordinator for the campus. She teaches graduate students in the department's three nationally accredited counseling programs and has served as the Counselor Education Department Chairperson and as Coordinator for the Student Affairs Professional Practice program. Some of her areas of professional focus have been ethics in counselor education, rural counseling practice, student affairs practice in higher education, college students and environments, professional roles and identity, ethical practice, cultural competence, and transition processes for students with disabilities. Dr. Burnell has been an active contributor to the design of nationally accredited graduate programs in teacher education and counselor education and coordinates the accreditation process for the Counselor Education Department. Dr. Burnell began her education career as a secondary mathematics teacher and worked for many years as a counselor in higher education settings.

Heidi L. Schnackenberg, PhD

Heidi Schnackenberg is a Professor in Educational Technology in SUNY Plattsburgh's (New York) Teacher Education Unit and is Coordinator of the Masters of Science in Education programs. She currently teaches both undergraduate and graduate classes on the use of technology to enhance teaching and learning in the P-12 classroom, social issues in education, ethical issues in educational technology, and childhood education courses as a member of one of the faculty cohorts in the BS program in Childhood Education. She previously taught in the graduate Educational Technology program at Concordia University is in Montreal, Quebec, Canada and has worked as an instructional designer for both Intel and Motorola corporations. Her various research interests include the integration of technology into pedagogical practices, the legal and ethical implications of western technologies in nonwestern and third-world cultures, and collaborative and cooperative learning strategies within student learning communities.

Dr. Schnackenberg recently designed and developed an extensive MyEducationLabs online supplement for the book *Foundations of American Education: Perspectives on Education in a Changing World* (15th ed.), published by Pearson Higher Education Publishers. She also co-edited (with Dr. Burnell) the 2012 book titled *Best Practices for Education Professionals*. Dr. Schnackenberg began her education career as an elementary music teacher.

CONTENTS

LIST OF CONTRIBUTORS

Aline R. Bobys, Ed. D.
SUNY Plattsburgh, New York, Email: bobysar@plattsburgh.edu

Beverly A. Burnell, PhD
NCC, State University of New York at Plattsburgh, New York, Email: burnelb@plattsburgh.edu

Caroline Knight, PhD
Coastal Carolina University, Conway, South Carolina, Email: cknight@coastal.edu
Jamia Thomas Richmond, PhD

Coastal Carolina
University, Conway, South Carolina, Email: jrichmon@coastal.edu

Claude M. Aldous, M.S.
SUNY Potsdam, New York, Email: aldouscm@potsdam.edu.

David Iasevoli, PhD
SUNY Plattsburgh, New York, Email: diase001@plattsburgh.edu

George Still, M.S.
SUNY Plattsburgh, New York, Email: gstil001@plattsburgh.edu

Maureen Squires, Ed. D.
SUNY Plattsburgh, New York, Email: msui001@plattsburgh.edu

Heidi L. Schnackenberg, PhD
State University of New York at Plattsburgh, New York, Email: heidi.schnackenberg@plattsburgh.edu.

Jelane A. Kennedy, Ed.D.
Professor, Counseling and College Student Services Administration, The College of Saint Rose, 432 Western Avenue, Albany 12203, New York 12203

Lauren T. Gonyea, M.S. Ed.
SUNY Plattsburgh, New York, Email: lgony001@plattsburgh.edu

Maureen E. Squires, Ed. D
SUNY Plattsburgh, New York, Email: msui001@plattsburgh.edu R. Luppicini, University of Ottawa, Canada, Email: rluppici@uOttawa.ca.

Stephanie Bennett, PhD
Associate Professor, Sociology, The College of Saint Rose, Albany 12203, New York

Wendy Neifeld Wheeler, PhD
Director of Parent and Family Relations, The College of Saint Rose, Albany 12203, New York

LIST OF ABBREVIATIONS

ACPA	American College Personnel Association
ADA	Americans with Disabilities Act
ADD	attention deficit disorder
ADHD	attention deficit hyperactivity disorder
ASHE	Association for the Study of Higher Education
CACREP	Council for Accreditation of Counseling & Related Educational Programs
CRTs	critical race theorists
GRE	revised general test
ICTs	Information and communications technologies
IDEIA	Individuals with Disabilities Education Improvement Act
LDS	learning disabilities
LGBTQ	lesbian, gay, bisexual, transgender, and questioning
MCCSAPI	Multicultural Competence Characteristics of Student Affairs Professionals Inventory
NASPA	National Association of Student Personnel Administrators
NCATE	National Council for Accreditation of Teacher Education
NCLB	No Child Left Behind Act
SAT	scholastic assessment test
SD	standard deviation
TE	technoethics
WWW	World Wide Web

PREFACE

The Ethics of Cultural Competence in Higher Education explores the concepts of ethics and culturally competent practices from the perspectives of higher education professionals in various roles (e.g. faculty, administrators and student services professionals), and includes the voices of both undergraduate and graduate students. The concepts of ethics and cultural competence can be as elusive as the words themself. Often, debates arise while considering what ethical behavior looks like or what sorts of actions (or thoughts!) are culturally competent. In reading and editing the various chapters that comprise this book, we have learned a great deal about each of these concepts, both from our contributing authors and from the discussions we have had with each other. As we talked about the ethics of leadership and the culturally competent practices of education professionals, we often found ourselves wandering around the gray area that comprises subjectivity with both of the topics. What may be considered ethical practices for one individual might be experienced as unjust by another. Also, similar arguments can be made with cultural competence. At the end of the day, though, we both realized that the goal is not an endpoint, where we have "arrived" as unquestionably ethical and entirely culturally competent. The goal is also not even the road one travels to get to either one of those places. In reality, the idea is to engage in the work, or start down the road, to examine ourselves and the ethical and culturally competent practices in our institutions and fields of study and to improve our own practices from whatever point we currently find ourselves.

In compiling the chapters in this book, our hopes and goals were to provide readers with thoughtful material that illustrates both the challenges and the rewards of striving toward ethical and culturally competent practice in higher education. We are clear that this work is aspirational since neither we nor any of us has reached that pinnacle of perfection in regard to ethical and culturally competent practice. Indeed, although we have edited a book on both of these concepts, we would never claim to be experts on either topic, nor are we pillars of either ethical behavior or cultural competence. But we are working at it. We are travelling our own journeys in our development of ethical values, sensitivity, and behaviors, and also in developing increasingly complex levels of culturally competent knowledge, awareness, and skills.

Our motivations for the creation of this volume were several. As professional educators who teach and have served in varied leadership roles in higher education settings, we both are committed to sound ethical and culturally competent practices in our work with students, colleagues, and members of the public with whom our graduates will work throughout their professional careers. A hallmark of a profession is that it is self-governing; that the members of the profession define both what constitutes ethical practice of the profession and the processes by which ethical practices are maintained so as to protect both the profession and the public (the users/consumers/recipients of the profession's services or products). In his 2013 article, "Northern Twilight: SUNY and the Decline of the Public Comprehensive College", Bob Golden contends that colleges and universities are not ivory towers, removed from the communities in which they reside. He argues that institutions of higher education need plenty of "social space" to operate effectively, and that university faculty should be trusted to act in the best interests of students, colleagues, and society. It stands to reason, then, that faculty members in higher education should be expected to act ethically and be culturally competent in their professional lives, and to engender these same traits in their students.

When our students graduate, they need to be prepared to work in diverse environments, but they are not being adequately prepared (Lee, Poch, Shaw, and Williams, 2012). Why is this? "Neither the institutional culture nor its participants and representatives (faculty, staff, administrators, students) are automatically willing or inherently competent to engage diversity effectively" (Lee et al., 2012, p. 5). It is our assertion that ethical practice in higher education requires ongoing development of cultural competence for both students and professionals. In making this assertion, we want to be inviting of all in higher education who want to be part of the conversation about what it means to be culturally competent and what it means for ourselves, our students, and the local, regional, and global communities in which we, our students, and our graduates participate outside of academia. We seek to engage all in higher education who want to contribute to, and learn from, the exchange of ideas about how institutional decision making, policy development, delivery of services, and, of course, pedagogy can become reflective of the diverse perspectives and full range of voices within our communities.

— Beverly A. Burnell and Heidi L. Schnackenberg

REFERENCES

Golden, R.; Northern twilight: SUNY and the decline of the public comprehensive college. *Thought & Action*, **2013,** *29,* 45–56.

Lee, A.; Poch, R.; Shaw, M.; and Williams, R. D.; Engaging diversity in undergraduate classrooms: a pedagogy for developing intercultural competence. *ASHE High. Educ. Rep.* **2012,** *38(2),* Retrieved December 19, 2013 from: wileyonlinelibrary.com. DOI: 10.1002/aehe.20002.

FOREWORD

My mind was racing while reading *The Ethics of Cultural Competence in Higher Education* (ECCHE). I appreciated the often, though necessary, repeated assertion that an overriding concern in higher education is to what extent faculty and professional staff demonstrate, and students would be able to acquire, a sophistication—cultural knowledge and cultural sensitivity—to more than adequately understand the plight of underrepresented individuals and groups. Due to the projected changing demography, "forewarned is forearmed" may be an appropriate adage. Though the future demographic should have never been the motivating factor for such self-development, instead, the motivating factor should always be "doing the right thing," and in the case of educators the right thing is being prepared to not just step into leadership moments when they arise, but to also know that you have done your best to be prepared for those moments should they arise.

However, just as importantly, the thought of whether or not any educator could actually become a "social justice" teacher without becoming an ally is a curious thought indeed. It was quite informative as well as validating to see the various authors in ECCHE relentlessly assert exactly that point. Yes, like so many other social justice advocates I'm in the camp that it is not possible to be an outstanding teacher of social justice without being an ally.

The scope of any collective discussion on the ethics of cultural competency in higher education should be broad, and this edited book doesn't disappoint. Beginning with a snapshot of various aspects of leadership, coeditor Heidi L. Schnackenberg's contribution ventures to many provocative places. Schnackenberg isn't shy about an expectation of academic leaders being transformative, equipped with "skill, experience, and an intuitive sense of people." She further asserts that far too often these qualities are not necessarily present and that unfortunately the "leadership finesse" to handle major initiatives is lacking. To enhance the possibility of success, an agreed upon vision of the collective journey must be arrived at by both the key stakeholders and their chosen leader. Schnackenberg expresses concern about inadequate processes and political machinations on college campuses (instead of student centeredness or the well-being of the academic institution itself) as motivating factors which ultimately contribute to inadequate leadership. Gender bias within the academy as a result of "otherness" is another aspect of leadership that Schnackenberg undertakes within her chapter. Ultimately, she makes the point that rounding up the usual suspects to adequately accomplish a task will only culminate in the possible adequate completion of the project, but not anything that could be considered noteworthy.

Rocci Luppicini's contribution to the book is an exploration of the relationship between educational technology the roles of instructors/students and university administration in creating a university culture. He articulates the very important reality of how different constituent groups come together to build meaning. Luppicini challenges the notion of "student-as-customer" as a possible threat to university culture.

Maureen E. Squires' chapter situates the reader to consider a reality in academia that is far too often marginalized in discourse about standards, procedures, or ethical practices in higher education. She asks how the rights of teacher candidates with disabilities are reconciled with professional expectations aligned with quality teaching, especially when those disabilities aren't limited to mobility limitations or orthopedic impairments. Squires does not hesitate to suggest that faculty understand their role in supporting students with special needs, and advocates that faculty proactively design academic supports while maintaining the integrity of the overall program. Furthermore, Squires consistently asserts throughout her chapter that "as a human service," education requires positively engaging other humans. She further takes time to frame outdated negative notions of disability that remain deeply ingrained in U.S. culture which serve as impediments that prevent faculty from becoming natural allies.

Lauren Gonyea uses anecdotes, references to film excerpts, and pertinent research to assert in her chapter that the concept of "cool," as a result of being an indicator of social inclusion/exclusion, reflects an invaluable dimension of social justice work. Gonyea, after defining "cool," diligently frames her point by articulating the phenomenon of cool across five different diversity themes, "socio-economic class," "race," "ability," "sexual orientation," and "gender." Essentially, Gonyea challenges educators to transcend their societal indoctrination and/or perspective on "cool." She asks that they confront problematic privilege and proactively engage it, thereby becoming "unequivocally cool."

Aline Bobys's chapter speaks specifically to the core theme of this edited book, the morality of the teaching enterprise. Beginning with her awe-inspiring receptiveness to relinquishing the traditional role as sole architect of her curriculum, continuing with her invitation that student and teacher act as equal participants in their daily interactions, Bobys models the method that she promotes, in essence walking her talk, creating a democratic classroom where the educational process transitions from "me to we." Bobys challenges her students and the reader to interpret literacy, due to globalization, as inseparable from a social, political, and cultural process. Furthermore, she shares with the reader how her students, in the process of becoming "literacy coaches," also come to understand that literacy education inclusively approached leads to them becoming considerate and proactive citizens.

Caroline Knight and Jamia Thomas Richmond's chapter uses interviews of ten college students to engage the merits of practical experience beyond the traditional diversity course often deemed adequate for social justice development of future educators. Along the way they debunk the myth of a culture of poverty and its inherent assumptions. David Iasevoli, in his chapter, challenges future teachers to reflect on their way of seeing, proactively engaging a society that far too often isn't egalitarian. Through strategic exercises designed to create conversations about oppressive language amongst other things, Iasevoli frames the pitfalls of social justice teaching within a monolithic educational environment.

Through the use of a "social constructionist conceptualization of masculinity," Claude Aldous and coeditor Beverly Burnell provocatively frame the necessity of higher education professionals to engage college men as men, and particularly men in transition, with various external stimuli continuously affecting their growth in every facet of their realities as members of a college community. Aldous and Burnell assert that college men aware of their gendered socialization and its accompanying privilege are better situated for proactive contributions to their academic community.

George Still and Maureen Squires provide results of an open-ended survey to engage the possibility of developing multicultural competencies by considering oneself and one's environment within the context of power and marginalization. In their chapter, Still and Squires speak at length about the reality of deficit-based perceptions of racially underrepresented children. They caution against a dominant group's cursory usage of multiculturalism as akin to "cultural tourism," and challenge pre-service teachers to become cross-culturally competent, essentially unpacking their own baggage while simultaneously learning about their students' realities.

Jelane A. Kennedy, Wendy Neifeld, and Stephanie Bennett's collaborative effort to assess cultural competency development for new student affairs professionals within the context of a social and cultural foundations course is somewhat of a forerunner to research that suggests one course isn't enough for future professionals. The authors of the chapter recognize the limitations of one course as well, but nonetheless proffer the merits of taking such a course for entry level practitioners as opposed to not taking it at all. More so, they frame the growth experienced by their student affairs graduate students as worthwhile. They articulate that, contrary to national challenges to do so, research revealed over a dozen years ago that there was a deficiency in cultural competency education for student affairs programs to the extent that only about 43% had a diversity course as part of the core curriculum.

All of the conversations within this book are invaluable to anyone seriously committed to a high quality and profound social justice effort. I have had the pleasure of directing two diversity and social justice centers for two well-respected universities, as well as having done significant work with highly touted and well-respected academic and business clients. Nonetheless, I read this book like a kid in a candy store hungry for a specific type of tasty treat while discovering other delicacies I had never previously considered. I highly encourage you, the reader, to take the time to read, consider, and ultimately act upon the lessons learned from *The Ethics of Cultural Competency in Higher Education*.

— J. W. Wiley, PhD,

Director, Center for Diversity,
Pluralism, and Inclusion, SUNY Plattsburgh, New York

CHAPTER 1

ASPECTS OF LEADERSHIP IN HIGHER EDUCATION: A SNAPSHOT

HEIDI L. SCHNACKENBERG

CONTENTS

> *Several years ago, I was a member of my university's search committee for a new Provost. During one interview, a candidate told the story of how she changed the college course schedule at the institution where she currently worked. She said that the previous schedule was essentially set at the convenience of the faculty and students often had great trouble getting the courses that they needed during the semesters in which they were expected to study a particular topic. Although appreciated by the student body, the candidate admitted that she made no friends among the faculty by instituting this change. In hindsight, she said, she would have worked harder to get more faculty buy-in before changing such a wide-ranging policy. But, she was firm that she didn't regret the initiative that she implemented and said that she would certainly do it again if a similar situation arose. The unfortunate outcome of this scenario is that the candidate could get no faculty support for any other initiatives that she tried to implement.*
>
> *After the candidate departed our campus, the search committee met to discuss her interview, qualifications, and how well she would fit as our new college Provost. As expected, the issue of the scheduling initiative came up as a topic of discussion. The committee went back and forth about the candidate's affective skills and her decision-making ability. The group was quite split as to whether or not they supported this candidate. At one point in the discussion, my colleague offered, "Regardless of whether you liked her or her decision, the fact is that what she did was the right thing to do. It was. It was right." To which I replied, "Leadership is more than just being right ..."*

In the 20 or so years that I have been in higher education, I have worked with many leaders, and even held minor leadership positions myself. I care very much about the people I serve, both students and faculty, and I often think about how I can do my job better. I reflect a lot on my own leadership abilities (or lack thereof) and the various challenges of administrative positions. What follows is a snapshot of my experiences with leadership, leaders, and the leadership process in higher education. It is not intended to be a comprehensive overview of the highpoints and pitfalls of administration, but rather a slice of a perception of leadership through the lens of a typical faculty member in higher education. I know that my reflections are piecemeal and rarely multidimensional, but they are indeed my experiences and I share them here.

To me, one of the interesting things about leadership in higher education is that these positions are usually filled with highly trained, very well-educated academics. Individuals who were prepared to be contributors of knowledge in their fields and/or teachers at the most advanced levels of schools and universities. It is fair to say that most of these men and women had much more educational training to become professionals in areas other than administration. In fairness, when some academics assume positions as department chairs, deans, vice presidents, provosts, etc., many of them are required to attend a certain amount of professional development on leadership skills. Of these, a few individuals even seek out and attend leadership academies to more deeply develop their skill set. But academics who truly dedicate the same amount of time to prepare for their positions of leadership as they did for their professorships are extremely rare. That is not to say that leaders in the academy do not try their best. On the contrary, I believe that most of them do. However, with underdeveloped skills, academic leaders often fail to successfully achieve many of the tasks they are responsible for accomplishing. Moreover, it is rare that they truly "lead" the people for whom they are held accountable. It is therefore incumbent upon the institution to choose their leaders wisely, and it is the responsibility of the chosen individual to hone and develop a professional skill set.

Much has been written about the transformative nature of leadership positions in education (Abbasi and Zamani-Miandashti, 2013; Argia and Ismail, 2013; Bogler, Caspi, and Roccas, 2013; Farhan, 2013; Leithwood, Jantzi, and Steinbach, 1999; Patterson, 2013; Wilson and Small, 2012). When individuals assume responsibilities in administrative positions, they are often asked to implement new initiatives, change group culture, create a shift in the way a group or institution does business, or some other large-scale task. While these are common administrative endeavors, to manage them successfully takes skill, experience, and an intuitive sense of people. These are abilities that academics in positions of leadership may or may not possess. Often, academic leaders are simply working on the basic competencies of good business managers—organization, availability, responsiveness, equity, and so on. Large-scale initiatives requiring leadership finesse are often far outside their scope of capability. And yet, university administration remains surprised when major initiatives are not successful. Why is this? Davies, Hides, and Casey (2001) posit that in their previous positions (generally as faculty), all that academic leaders had to focus on was the curriculum and student developmental needs. In more recent times, with institutions of higher education adopting more of a business model, academic leaders have had to assume the role of managers, supervisors, counselors, facilitators, and sometimes remain as part-time teaching faculty, with all of the attached project management, institutional knowledge, and people skills. This juxtaposition of academic preparation and administrative role is one that at present, often does not work in the academy.

Knight (2001) contends that leadership and management are entirely different practices. He states that leadership is inextricably connected to change, whereas management is simply an efficient perpetuation of the status quo. In my observations, I have seen academic faculty most often want individuals to guide and support them, and administration want individuals who can implement initiatives linked to change. Both parties then want leaders more than managers. Unfortunately, the "system" of higher education appears to want to operate as a business, which requires managers rather than individuals who are agents of change (i.e., leaders). This conflict of roles and responsibilities frequently leaves academics, some of whom are unprepared for their leadership roles to begin with, fully at odds with what their various constituencies want and expect of them. This type of confusion about the type of leader that is needed and/or wanted by a department, division, unit, etc., further adds to the potential failure of an academic administrator.

Clearly, one of the first steps to ensure the success of an academic leader is for the individual and the group that they serve to collectively define the position in which the administrator will serve. While this may seem simple, honoring the various desires and needs of all the individuals and groups that the leader will answer to is a difficult one. However, using methods of self-governance and democratic processes, a group can come to consensus about what it expects of their leader. This type of collaborative determination of what a position is supposed to be helps define the expectations for the individual who will fill it. This then aids in the process of finding the "right" person for the job. Grummel, Devine, and Lynch (2009) argue that how the leadership role is constructed, in large part, determines the selection process. Unfortunately, simply finding "the best fit" is not always the only factor at play when higher education chooses and appoints administrators.

The institutional process by which academic leaders are chosen is often flawed, and fraught with political considerations, rather than the best interests of the students, the degree programs, the faculty, and the institution. Lower-level administrative positions (e.g. department chairs) are often either elected, assigned by the dean, or assumed by default (because there may not be any other viable candidates to take the position, or it just may be someone's "turn" to do the job). Although not the most prestigious positions in the institution, and at times even viewed as paper-pushing jobs, department chairs are often the first individuals that students or faculty go to when they have questions, challenges, complaints, etc. These positions are extremely important because when constituents get their needs met or their issues resolved by a department chair, it leaves administrators in higher positions to handle other, more complex issues. It is essentially the way the chain of leadership is intended to work. However, if department chairs are not equipped, or even truly interested, in doing their jobs well, then faculty and students will get their needs met and voices heard in other places (i.e., the dean, the union, etc.). Although their roles are often viewed as unglamorous, department chairs are extremely important to the day-to-day functioning of colleges and universities, and it would do well for institutions to invest in the individuals holding these positions and invest some time in choosing the best individuals for the jobs.

Unlike department chairs, associate deans, deans, provosts, etc., are often vetted and selected by search committees (with the next highest administrator being the hiring authority); but at times, they are also appointed by the dean, the provost, etc., him/herself. Often the individuals selected into these positions have previously been department chairs, deans, etc., and have some knowledge about the workings of the institution and training as some type of administrator. Therefore, they have had some professional development and/or experience to bring to a more complex position. Search committees do their best to vet the candidates since positions of this nature are often well-defined by the needs of the dean's office, provost's office, president's office, or whichever department in which they will work. However, on some occasions, these individuals are appointed by higher administration, thereby subverting the search process and devaluing faculty input as to what they think a department, division, or institution may need. This type of appointment rather than selection often does the most disservice to the individual holding the position since the faculty do not truly know his/her skill set, experience, accomplishments, or even to what extent they can trust the person to make decisions in the best interest of the group with which they are to work. Husu (2000) adds that various biases are likely to occur when the evaluation process [of potential administrators] is kept confidential. Unfortunately, by acting outside of accepted practices and processes, administration sends the message to the faculty that they either do not care what they think, do not trust their ability to choose who leads them, or do not have the capacity to know what's best for the institution. The faculty then receives this message, directly or indirectly, and become insulted and sometimes resentful. In general, the focus of faculty members' subsequent negative feelings is the individual who was appointed to the administrative position. By ignoring approved processes, administration has already set the individual, the faculty, and any other stakeholders, at a disadvantage before a working relationship has even begun. If the individual had been selected for the position by his/her future colleagues, based on merit, it would cast a more positive atmosphere on the start of the new working relationship.

Discussing the use and/or misuse of institutional procedures and processes illustrates the importance of what is likely the most key attribute that any leader can possess—honesty. It is often the mistaken idea among administrators, managers, etc., that they need to appear infallible, without the need for counsel from others, and in possession of the ability to make the right decisions in every situation. In reality, this type of leader simply does not exist and the people for whom s/he is responsible generally realize that. It is interesting, therefore, that often faculty are told by administrators only semi-accurate versions of situations, or partial truths about issues, initiatives and challenges, under the guise of making difficult challenges easier to understand and accept, while all the while administration is going through the motions to mask their own inadequacies (and/or dishonest dealings), which faculty members realize exist in the first place. All of the effort undertaken by leaders to create scenarios that may be only partially truthful is, in fact, wasted machinations. In almost every scenario in higher education, faculty members simply want the facts of a situation, as unglamorous as they may be, in order to genuinely assess a problem. By not being honest with the people whom they oversee, administrators engender lack of faith and trust in their abilities and leadership. While this reaction is warranted, it is also harmful for the progress of the institution. The faculty has the capacity to make informed decisions, without being patronized by their leaders, and at the same time can accept administrative decisions that are made openly and honestly.

Interestingly, the issues of professional preparation, dispositional characteristics, selection processes, and job descriptions all play strongly into a gender bias that exists in higher education administration. Morley (2013) notes that there has long been and still remains an under-representation of women in senior administrative positions in higher education. She states that this gender imbalance means that women are not part of the decision-making process in a variety of venues, including committees, administrative boards, and executive councils, thereby omitting critical perspectives from a large portion of the population of the college. Gronn and Lacey (2006) theorizes that gender bias often occurs when the dominant group "clone" themselves by appointing individuals such as themselves to minimize risk of change from new voices pursuing fresh perspectives, ideas, and agendas. This practice often leads to the appointment of administrators, rather than a more open selection search process facilitated by a committee. Rees (2010) argues that transparency in the selection process can benefit women, and indeed likely other under-represented individuals. It stands to reason then that the converse is also true; top-down appointments of administrators do not favor individuals from nondominant groups.

The lack of women in leadership positions has long been blamed on women's domestic responsibilities and the demands of motherhood. However, this does not explain the dearth of single, child-free women in higher education administration (Currie, Thiele, and Harris, 2002). It has been my experience that male administrators often do not know "what to make of" strong women in leadership positions and often need to "slot" them into specific categories in order to know how to "deal with them." Women administrators are judged to be aggressive tigresses, humble placators, peace-makers, nurturers, teases, flirts, or a variety of other characters. Women leaders in higher education cannot simply just do their jobs as the people that they are because they are still viewed as the "other." Probert (2005) posits that women must manage this otherness and minimize their gender difference in order to be successful in higher education. It is stunning to reflect that women leaders cannot be both feminine and in charge in the present-day work world. These extra challenges that women face (and men do not) when considering administrative positions often deters females from pursuing leadership posts (Kram and McCollum-Hampton, 2003). Indeed, in the institution where I currently work, there is only one female in upper administration and only one, new, female dean. Sadly, in the past decade, there have only been four or five women in upper administration in total. Alas, it appears that the female voice has been silenced at my own institution as well.

In closing, I must wonder why the various phenomena that exist within higher education, and that I discuss here, keep perpetuating themselves, particularly if they have been found to be problematic for a very long time. Administrators have been and continue to be culled from the ranks of academics without any forward thinking or consideration of training individuals interested in positions of leadership. Ongoing professional development is rarely offered or sought once leaders attain their positions, partly due to neglect and preoccupation with other initiatives, and partly due to a de-prioritizing of continuing education for faculty and administrators. The search versus appointment duality of processes continues to erode the faith of the constituents in the administrators in higher education. Also, women and minorities continue to suffer from this lack of transparency in the decision-making process. The duplicity [of not respecting and following institutional processes] engenders distrust, lack of communication, and destruction of the very community [leaders are expected] to build (B. Burnell, personal communication, July 16, 2013). Honesty continues to be a cornerstone of good leadership, but it is most often utilized haphazardly at best. Without this cornerstone, leadership then stands on a weak foundation. Is this truly what we want for the academy? Albert Einstein is generally credited as saying that the definition of insanity is doing the same thing over and over again and expecting a different result. Perhaps, it is time that something changes in higher education in the way we choose and prepare our institutional leaders.

KEYWORDS

- **Higher education**
- **Leadership abilities**

REFERENCES

Abbasi, E.; and Zamani-Miandashti, N.; The role of transformational leadership, organizational culture and organizational learning in improving the performance of Iranian agricultural faculties. *High. Educ.*, February **2013,** 1–15.

Argia, H. A. A.; and Ismail, A.; The influence of transformational leadership on the level of TQM implementation in the higher education sector. *High. Educ. Stud.*, **2013,** *13(1),* 136.

Bogler, R.; Caspi, A.; and Roccas, S.; Transformational and passive leadership: an initial investigation of university instructors as leaders in a virtual learning environment. *Educ. Manage. Admin. Leadership.* **2013,** *41(3),* 372–392.

Currie, J.; Thiele, B.; and Harris, P.; Gendered Universities in Globalized Economies: Power, Careers, and Sacrifice. Lexington, KY: Lexington Books; **2002**.

Davies, J.; Hides, M. T.; and Casey, S.; Leadership in higher education. *Total Qual. Manage.* **2001,** *12(7–8),* 1025–1030.

Farhan, B. Y.; Demand-driven higher education: entrepreneurial behaviour and the role of transformational leadership. Retrieved June 12, 2013 at Social Science Research Network: http://papers.ssrn.com/sol3/papers.cfm?abstract_id=2197576, **2013**.

Gronn, P.; and Lacey, K.; Cloning their own: aspirant principals and the school-based selection game. *Aus. J. Edu.* **2006,** *50(2),* 102–121.

Grummel, B.; Devine, D.; and Lynch, K.; Appointing senior managers in education: homosociability, local logics, and authenticity in the selection process. *Educ. Manage. Admin. Leadership.* **2009,** *37(3),* 329–349.

Husu, L.; Gender discrimination in the promised land of gender equality. *High. Educ. Eur.* **2000,** *XXV(2),* 221–228.

Kram, K.; and McCollum-Hampton, M.; When women lead: the visibility-vulnerability spiral. In: Ely, R.; Foldy, E. G.; and Scully, M.; Eds., Reader in Gender, Work, and Organization. Oxford, UK: Blackwell Publishing; **2003,** 211–223 p.

Knight, P.; Departmental Leadership in Higher Education. Philadelphia, PA: Open University Press; **2001**.

Leithwood, K.; Jantzi, D.; and Steinbach, R.; Changing Leadership for Changing Times. Florence, KY: Taylor & Francis; **1999**.

Patterson, S. M.; Layers of critical engagement: exploring the intersections of leadership, critical theory, and learning. *About Campus.* **2013,** *18(1),* 9–15.

Probert, B.; "I just could not fit it in": gender and unequal outcomes in academic careers. *Gender, Work, & Organ.* **2005,** *12(1),* 50–72.

Rees, T.; In the last three years I have been a Pro Vice Chancellor .... In: Women in Engineering and Technology Research. Godfroy–Genin, A. S.; Ed. Berlin: Lit Verlag; **2010,** 21–29 p.

Wilson, E.; and Small, J.; TEFI 6, June 28–30, 2012, Milan, Italy: Transformational leadership for tourism education. *J. Teach. Trav. Tourism.* **2012,** *13(2),* 202–206.

CHAPTER 2

TECHNOETHICAL REFLECTIONS ON THE EVOLVING ORGANIZATIONAL CULTURE OF UNIVERSITY 2.0

ROCCI LUPPICINI

CONTENTS

2.1 INTRODUCTION

I have spent the larger part of my academic career tracing out social and ethical meaning of technologies (technoethics) within institutions and society. One particularly interesting area of inquiry connected to educational technology and research in technoethics focuses on social and ethical intertwinement of educational technology within university culture. This intertwinement applies to considerations of technology both within and beyond the classroom.

2.1.1 TECHNOLOGY IN THE CLASSROOM

Technology in the classroom has been around as long as chalk and chalkboards. When one speaks of technology in the classrooms today, however, it is the use of digital technologies that comes to mind: multimedia LCD systems, smart boards, clicker technology, tablets, and the one technology that is most common sight in university cafes, cafeterias, and classrooms around the continent, and that is the laptop computer. Laptop computers were seen as so central to student learning that some private schools and some universities (i.e., Wakefield, Acadia University) required laptops to each enrolled student as part of their tuition fee (Luppicini, 2010). Almost all universities I am familiar with now have wireless Internet access for all university members anywhere on campus and encourage instructors to embrace new technologies in classroom teaching and learning. But herein lies the paradox. On the one hand, electronic technologies can be used by students and instructors within university classrooms to advance research speed and capacity, improve note taking, and expand collaborative learning possibility and much more. Having students break off into smaller groups to discuss assigned class tasks in online chat rooms or use Google Docs to collaboratively work on a class assignment can really improve classroom engagement. This is especially salient in large 200+ classrooms where discussion is difficult. On the other hand, misuses of electronic technology within organizational settings such as cyber loafing, network attacks, information theft, unauthorized transmission of confidential information, cyber bullying, and the exchange of online pornography are serious problems in contemporary organizational life that have trickled into educational settings. The misuse of electronic technology in the classroom can promote distraction (Bugeja, 2007), low-level learning due to multitasking (Hembrooke and Gay, 2003), and classroom disruptions. In some cases, instructors have enforced bans on electronic technologies in the classroom as a countermeasure (Cole, 2008). This gives rise to an ethical question of values that is not easy to deal with: When faced with electronic technology abuse in the classroom (and/or competing with technology for student attention that can detract from learning), what are the ethical considerations and possible consequences of typical measures taken? (e.g. restricted use policies, instructor imposed classroom bans, university imposed bans, no action, etc.). This question gets at the challenge of balancing the university culture, values, and responsibilities with the individual rights and freedoms of culture members (students and professors).

2.1.2 TECHNOLOGY BEYOND THE CLASSROOM

In the early to mid-2,000s, I became intrigued by the potential of communities of learning and practice (Lave and Wenger, 1991), moving online and how this would contribute or detract from education and training. In an edited volume titled *Online Learning Communities* (Luppicini, 2007), educational researchers and instructors focused on a number of key areas of online learning community research, including online community types, instructional design considerations for online learning communities, applications of online learning communities to enhance student collaboration, and international perspectives on online learning communities. This work traces the popularization of online learning in universities and put into perspective key opportunities for designing, developing, and implementing online learning communities to augment traditional classroom instruction.

The naturalization of online learning into the teaching and learning culture of universities also had a number of associated challenges that revealed a dark side that went beyond the university walls. Easy online access to and dissemination of personal student information led to serious problems such as cyber stalking, cyber bullying, information theft, identity theft, and online plagiarism (See Luppicini, 2009 for a review.) This gives rise to ethical questions of value that are not easy to deal with: When faced with electronic technology abuse outside the classroom, what are the ethical considerations and possible consequences of typical measures taken by university administration and faculty? (e.g. rules of netiquette, use of online plagiarism detectors, public computer monitoring, legal services for student and faculty for cybercrimes, etc.). This also gets at the challenge of balancing the university culture, values, and responsibilities with the individual rights and freedoms of culture members (students and professors).

Recent work appearing in a co-edited volume titled *Education for a Digital World: Present Realities and Future Possibilities* (Luppicini and Haghi, 2012) takes a looks at organizational and sociocultural aspects of higher education. It explores student online anonymity, privacy practices (Luppicini and Lin, 2012), religious and cultural values in the digital age (Luppicini and Myhill, 2012), and online and offline student identity practices (Luppicini and Barber, 2012). Taken together, this recent work looks beyond everyday teaching and learning to provide a glimpse at the complex intertwinement of values, identity, and practices of students living within the technological culture of University 2.0. In an effort to shed additional light on the ethical complexity of this evolving organizational, this chapter applies a reflective lens guided by the research field of technoethics to explore the values and roles of instructors, students, and university administration at University 2.0. The next section provides key background information on technoethics and educational technoethics. Following this, the chapter provides a reflection on key ways in which the advancement of technology is challenging conceptions of organizational culture roles and values in higher education. The chapter contributes to the volume by raising awareness about the close connection between educational technology, human values, and university culture.

2.2 BACKGROUND

2.2.1 TECHNOETHICS AND EDUCATIONAL TECHNOETHICS

Within a society where technology progresses more rapidly than our ability to master it, social and ethical consequences have become important considerations in technological development and its integration into society. As a result, concerted efforts to study the ethical aspects of technology in life and society have been nurtured in the field of research known as technoethics. The term *technoethics* was first used by the famous philosopher of science, Mario Bunge in the 1970s (Bunge, 1977) when arguing for greater moral and social responsibility among technologists and engineers concerning their creations. This concern for ethical aspects of technology evolved into a field of study under this same name. According to the *Handbook of Research on Technoethics*:

> Technoethics (TE) is concerned with all social and ethical aspects of the design, development, utilization, management, and evaluation of science and technology in society. As an interdisciplinary field, it utilizes theories and methods from multiple knowledge domains to provide insights on ethical dimensions of technological systems and practices for advancing a technological society. (Luppicini and Adell, 2009)

Technoethics has been applied to a number of fields including computer science, engineering, communications, biomedicine, nanotechnology, and education. Within the realm of education, technoethics in education (educational technoethics) begins with the assumption that the key to understanding technology and its impact on education looks beyond technical considerations in an effort to examine how technology links to educational culture and values.

Educational technoethics represents the area of technoethics that deals with ethical issues associated with technology and education (Cortés, 2005). There are many topics covered including online plagiarism, online diploma mills, cyber-bullying, cyber democracy, digital divide, e-learning ethics, emancipatory educational technology, professional technoethics, and technoethical assessment. Key pioneering studies in this area include contributions from Nichols (1987; 1994) and Boyd (1991), which explore negative aspects of educational technology, while highlighting the need for emancipatory uses of educational technology as a means to advancing ethical guidelines and the role of human values when approaching the new technologies for educational aims.

2.2.2 CULTURE AND THE TURN TO NEW MEDIA

What is culture? The meaning of the term "culture" may best be described as an evolving concept to accommodate the changing configurations of society and how people live with one another, build meaning together, and define themselves within their respective communities. Traditionally, culture was seen as the collection of symbols, language, behaviors, practices, customs, traditions, beliefs, and values held by a group of people brought together with shared needs and common challenges in life (Geertz, 1973). When one thinks of the term "culture," it may bring to mind a group of people living together in a community defined by their history, rituals, practices, language spoken, and geographic location. Certainly, this was true one hundred years ago. In anthropology, a typical ethnographic study of a culture would mean travelling to some remote area and immersing oneself within a settlement or village for field research on the local cultural practices (aboriginals, Zulu, Amazonians, etc.). But then, the world became global and how people came together to build meaning changed.

The notion of a global culture (and acknowledgment of the role that media played in it) was a game changer in how the term "culture" was viewed. In the early 1960s, Marshall McLuhan traced the shift in society from an individualistic visual print culture to an oral culture (tribal culture) driven by electronic media, greater interdependence, and a sense of collective identity that transcended geographical boundaries. McLuhan (1962) referred to this new type of social and cultural organization as the "global village":

> Print is the extreme phase of alphabet culture that detribalizes or decollectivizes man in the first instance. Print raises the visual features of alphabet to highest intensity of definition. Thus print carries the individuating power of the phonetic alphabet much further than manuscript culture could ever do. Print is the technology of individualism. If men decided to modify this visual technology by an electric technology, individualism would also be modified. (p. 158)Viewing the world as a global village connected through media brought both new meaning and complications, not only in how the term "culture" was viewed but also how culture was pursued. McLuhan and Powers' *The Global Village: Transformations in World Life and Media in the Twenty-First Century* (1989) applied the idea of the global village to document the new complexities in contemporary life and culture now subject to multiple cultural mindsets; that is, visual (linear and individualistic) and acoustic (nonlinear and many-centered) oriented cultures. In this sense, the global culture of media acted as a cultural glue uniting all cultures within a global culture anchored in media communications. In other words, it changed the view of culture as "over there" to culture as "here and everywhere."

This liberal take on culture also acted as a catalyst, opening the door to other areas where culture was being studied, namely, within organizations (including universities) where professional cultures were beginning to become the focus of inquiry (Trice and Beyer, 1993). Part of the reason can be attributed to the fact that individuals spend significant amounts of time working within organizations and engaged in professional practices that align with specific professional associations, professional practices, and professional values and codes of conduct/responsibility sanctioned. Thus, why not look at them along cultural dimensions as well as they do unite a group of people and this grouping can create shared values, practices, and rituals that become entrenched over time. This is particularly salient in higher education where professors and researchers come together within universities to pursue the advancement of knowledge for (the greater good of) society.

2.2.2.1 FROM WEB 2.0 TO UNIVERSITY 2.0

What is Web 2.0 and what is the relevance of this concept to our understanding of universities? Originally, the term "Web 2.0" was used by the information architect Darcy DiNucci, to describe the evolution of the World Wide Web (WWW) from a medium which permitted the transfer of information from one to many in a static online environment to one which promoted a higher degree of social interactivity and the creation of user-generated content using social networking applications. As stated by DiNucci (1999):

> The Web we know now, which loads into a browser window in essentially static screenfuls, is only an embryo of the Web to come. The first glimmerings of Web 2.0 are beginning to appear, and we are just starting to see how that embryo might develop. The Web will be understood not as screenfuls of text and graphics but as a transport mechanism, the ether through which interactivity happens. It will [...] appear on your computer screen, [...] on your TV set [...] your car dashboard [...] your cell phone [...] hand-held game machines [...] maybe even your microwave oven (.32).

The relationship between the early Web 1.0 to Web 2.0. is mirrored in the evolution of the university in that university education today is no longer limited to campus-based knowledge and skills acquisition that takes place within university classrooms. Based on this definition of Web 2.0, this author defines University 2.0 as a multifaceted place of higher education learning characterized by active student engagement with the Internet and digital media in conjunction with traditional classroom teaching that requires the physical co-presence of teacher and students. Learning management systems, online tutorials, simulated learning environments, virtual communities of learning, and online group work provide added value to traditional university teaching and learning practices that preceded the Internet age.

Within University 2.0, the incorporation of the Internet and multimedia technologies into university teaching offers more flexibility, greater access, and opportunities for collaborative learning not possible before the Internet. The advantages of University 2.0 are particularly evident within large undergraduate classrooms where new opportunities for student–student and student–teacher interaction are made possible by digital technology use. It is this vision of the university as University 2.0 with which this chapter aligns itself.

2.2.2.2 TECHNOETHICAL REFLECTIONS ON THE CHANGING CULTURE OF UNIVERSITY 2.0

What does culture mean in a university context, and where did it start? One can argue that the history of academia is also a history of the culture and values of intellectuals (and intellectuals in training) within society, upholding certain values, working toward certain goals, and engaging in certain practices and rituals. The most common place of these intellectual cultural practices, the "university," a term derived from the Latin (universitas magistrorum et scholarium), roughly translates as "community of teachers and scholars."

Stemming from earlier biblical use, the term "ivory tower" has been used in modern society to label universities where there is a strong culture of intellectual engagement that prioritizes the pure pursuit of knowledge, often disconnected from the practical concerns of everyday life. Examples include Oxford University (UK), La Sorbonne (France), and Harvard University (USA). It is no coincidence that in the United States, the term "ivy league" is still used to designate traditional institutes of higher education (Brown University, Columbia University, Cornell University, Dartmouth

College, Harvard University, Princeton University, the University of Pennsylvania, and Yale University), that emphasize academic excellence in the pursuit of knowledge, high admission standards, and sometimes snobbery and social elitism (Epstein, 2003).

In 2013, the ivory tower has fallen (for most of us). In the midst of tough economic conditions, and increasing competition for limited resources, universities have been confronted more and more with the need to adapt their image and roles to meet the interests (often conflicting) of key stakeholders (students, professors, alumni, funding agencies, government, corporate sponsors, etc.). This is a serious challenge since stakeholder interests can be in conflict with each other, and the core values of universities to uphold excellence in academic research and teaching. This includes the acceptance of donations and engagement in university–corporate and university–government partnerships. It is also an important precondition to a living university culture (and all cultures) because it focuses on what is needed for a culture to survive and flourish within the larger environment within which it exists. In this case, attention to the conditions for university (and university culture) survival in the current economic environment begs the question: How do universities stay alive and avoid selling out their culture(s)?

2.2.2.3 WHAT IS THE ROLE OF THE UNIVERSITY 2.0?

How do universities maintain their values as a culture of learning and knowledge advancement while getting an edge (or staying alive) in the competitive marketplace where emerging technologies, marketable skills, and innovation are hallmarks of contemporary economy and technological culture (Luppicini, 2010)? Digital technology plays a significant part in helping universities define themselves and their roles in terms of their teaching and recruiting strategies. The motto for university where I obtained my PhD was, in fact, "Real education for the real world." Their approach was to appeal to the working-class student by highlighting the practical focus for the job market and catering to part-time and mature students by offering many night classes, flexible programs, and of course, online course options. While distance education has been around in its contemporary form since the 1970s (see Luppicini, 2008), traditional universities have since capitalized to attract students by offering more flexible course options that can be pursued online and count toward their degree. Other universities offer high-tech training foci to keep up with the technological society we live in (e.g. MIT, Virgina Tech, Georga Tech, Ontario Institute for Science and Technology).

Second, many universities have a strong online presence on YouTube and Facebook to advertise and attract students. My own university presents itself on YouTube as Canada's university and the "center of national dialogue":

> Designed to reflect the excellence of our institution, the video highlights three of the University's exceptional strengths—a center of national dialogue, a research and policy leader, and a key international player. (University of Ottawa, 2013)

As can be seen in the above-mentioned examples, the technological culture and values of universities are discernable in the teaching, branding, and advertising of the university culture and what it offers. Universities actively engage in branding strategies to attract the best quality students, professors, and researchers, (along with the needed financial support required) for carrying out teaching and research activities. Following the lead of private sector organizations, contemporary universities now actively use branding strategies and online advertising to broadcast their services and personality (specializations), made distinctive by their positioning and unique culture (Hankinson and Cowking, 1993).

2.2.2.4 WHAT IS THE ROLE OF PROFESSOR 2.0?

Students' opinions of professors (and their teaching) influence tenure, and this influences the university culture in terms of its professor member status and retention. With so much on the line for new tenure-track faculty, student course evaluations are key considerations at most, if not all, universities. What complicates matters is the role that technology can play in promoting and discrediting the tenure-track faculty. One good example is, RateMyProfessors.com (RMP), a popular review site that allows college and university students to rate professors and campuses in Canada, the United States, and the United Kingdom. Professors are ranked on easiness, clarity, helpfulness, interest, and hotness (Ratemyprofessors, 2013). Many of my colleagues have their own professional websites and Facebook profiles (with varying degrees of personal information disclosed), and Twitter feeds that allow their students to follow their professional and personal activities.

In an effort to personalize (and humanize) their role as professors within the larger university culture, the traditional boundary separating professor from student has become blurred to some extent. And it poses the following questions: We are often encouraged to be approachable to our students, but is it acceptable to interact with students outside the classroom and, if so, in what contexts? And what types of online exchanges are appropriate and inappropriate to have with students? The recent trend in sexting (sending sexually explicit messages and/or photographs between computer and mobile phone users) is a primary example of the inappropriate use of information and communications technologies (ICTs), which can detract from the legitimacy and value of the educational use of ICT's to leverage appropriate teacher–student and student–student interactions for educational purposes (see Lane, 2011). Given that students' and professors' use of online communications and online interactions range from resume posting to sexting, the question of appropriate conduct offline also applies to professor conduct online as well.

2.2.2.5 WHAT IS THE ROLE OF STUDENTS?

University students are seen as learners who attend universities with the objective (typically) of receiving post-secondary instruction. This much may still be true, but who are the students in the eyes of university administration and professors? At University 2.0., technology is one of the key ways that students (customers) locate their institution of higher learning, and interact with it once they are enrolled. In this way, technology has entirely changed the way students learn about and assimilate into university culture. For instance, learning management systems are commonly used in University 2.0 to organize and track student activity, manage student assignments and marking, and deliver course content (EDUCAUSE Evolving Technologies Committee, 2003). This is just one of the many technologies helping to reshape university teaching and learning activities.

At University 2.0, students can be in contact with their parents and extended family anytime and anywhere on their cell phones. Facebook and Twitter provide additional ways for parents to stay in contact and track their children (and their professors). The term "helicopter parent" is increasingly used in university contexts to describe parents complaining to professors about student grades received (Kelley, 2008), often using the high cost of tuition and their entitlement as "customers" to justify their interventions (Alsop, 2008). It is not surprising that university administrations have adopted the increasingly popular habit of calling the university students "customers" (Svensson and Wood, 2007). While this may not be politically or legally contentious, it does have ramifications in terms of social and cultural meaning that raise the question: What are the ethical considerations connected to the changing conceptualization of students (customers) within University 2.0? And how might this influence university culture and the ethical treatment of students (and faculty members)?

In my view, the students-as-customers-type terminology threatens university culture in two ways. First, it risks reducing the meaning of universities and what they represent to a commercial market of exchange between sellers and customers. Instead of highlighting the changing direction of research and teaching focus to meet the needs of the real world, it highlights the university as an anonymous place of market exchange independent of the distinctive roles that professors and students have as shapers of university culture. Second, it highlights the status of students as external to university operations, decision-making (and meaning creation), and separate from the university. Personally, I see student participation in university culture-building as essential and many of my best insights and research collaborations have been in collaboration with my students. In my view, the convenient student-as-customer perspective runs counter to the organic nature of learning and exchange that is a hallmark within any university culture (traditional and contemporary).

2.3 CONCLUSION

The battle over technology in higher education is over. It is not that technology has won and humans have lost (technology 1: humans: 0). Rather, we have realized that technology is part of us (We did create it after all!) And what we are battling is how to best use that extension of the human self and society to leverage the satisfaction of our needs and interests, in this case, education. In terms of offering practical suggestions on the advancement of university culture from a technoethical perspective, there is no more important lesson to take away than the idea that culture begins at home, in this case, the classroom, which is the focal point of university life and where a solid culture of collaborative learners is built from the ground up. One possible strategy is to look closely, not at the learning objectives or technology but, rather, at the conditions underlying effective technology use. Luppicini (2010) advanced a set of conditions for effective technology use in the classroom to help guide culture-enabling strategies:

(i) Good technological practices that encourage contacts between students and faculty are conditional upon the presence of appropriate organizational strategies and rules of netiquette.

(ii) Good technological practices that foster reciprocity and cooperation among students are conditional upon the establishment of a learning community or participatory social network within an explicit learning context.

(iii) Good technological practices that foster active learning among students are conditional upon effective instructor strategies to actively engage students in their learning.

(iv) Technological practice promoting effective student–student and instructor–student feedback is conditional upon how instructors regulate the use of available feedback mechanisms.

(v) Good technological practices promoting time on task are conditional upon teaching strategies encouraging time on task technology use and discouraging time off-task technology use.

(vi) Good technological practices that fulfill high teaching and learning expectations are conditional upon the understanding of learning aims and objectives driving teaching and learning expectations.

(vii) Good technological practices that promote respect for learning diversity are conditional upon effective instructional strategies that instill respect for learning diversity.

These strategies originally intended for leveraging effective technology use in the classroom can also be adapted to university culture building for students and faculty by helping them organize around shared values and goals that are core to educational aims and university culture.

KEYWORDS

- **Culture**
- **Higher education**
- **Technoethics**
- **University 2.0**

REFERENCES

Alsop, R.; The Trophy Kids Grow Up: How the Millennial Generation is Shaking Up the Workplace. Jossey-Bass; **2008**.

Boyd, G.; The shaping of educational technology by cultural politics, and vice versa. *Educ. Train. Technol. Int.,* **1991,** *28(3),* 87–96.

Bugeja, M.; *Distractions in a wireless classroom.* The chronicle of Higher Education. Retrieved August 1, 2013, from http://fp.arizona.edu/geog695c/PDFs/wireless.classroom.pdf, **2007**.

Bunge, M.; Towards a technoethics. *Monist.* **1977,** *60(1),* 96–107.

Cole, D.; Why I ban Laptops in my Classroom. *Encyclopedia Britannica Blog.* Retrieved November 20, 2008 from http://www.britannica.com/blogs/ 2008/10/why-i-ban-laptops-in-my-classroom/, October 23, **2008**.

Cortés, P. A.; Educational technology as a means to an end. *Educ. Technol. Rev.* **2005,** *13*(1), 73–90.

DiNucci, D.; Fragmented future. *Print.* **1999,** *53*(4), 32–33.

EDUCAUSE Evolving Technologies Committee. Course Management Systems (CMS). Retrieved November 15, 2013, from http://www.educause.edu/ir/library/pdf/DEC0302.pdf, **2003**.

Epstein, J.; Snobbery: The American Version. Houghton Mifflin; **2003**.

Geertz, C.; The Interpretation of Cultures: Selected Essays. New York; **1973**.

Hankinson, G.; and Cowking, P.; Branding in Action: Cases and Strategies for Profitable Brand Management. London, England: McGraw-Hill; **1993**.

Hembrooke, H.; and Gay, G.; The laptop and the lecture: the effects of multitasking in learning environments. *J. Comput. High. Educ.* **2003,** *15*, 46–64.

http://www.ratemyprofessors.com/search.jsp?query=luppicini&queryoption=TEACHER&prerelease=true&search_submit=Search

Kelley, T.; Dear Parents: Please Relax, It Is Just Camp. The New York Times. The New York Times Company; July 6, **2008**.

Lane, F.; Cybertraps for the Young. Chicago: NTI Upstream; **2011**.

Lave, J.; and Wenger, E.; Situated Learning: Legitimate Peripheral Participation. Cambridge University Press; **1991**.

Luppicini, R.; Review of computer mediated communication research for education. *Instruct. Sci.* **2006,** *35*, 141–185.

Luppicini, R.; Trends in Educational Technology and Distance Education in Canada. Berlin: VSM Publishing; **2008**.

Luppicini, R.; Ed. Online Learning Communities. Charlotte, NC: Information Age Publishing; **2007**.

Luppicini, R.; Introducing Technoethics. In: Handbook of Research on Technoethics: Volume I. Eds. Luppicini, R.; and Adell, R.; Hershey: Idea Group Publishing; **2009**.

Luppicini, R.; Technoethics and the Evolving Knowledge Society: Ethical Issues in Technological Design, Research, Development, and Innovation. Idea Group Publishing; **2010**.

Luppicini, R.; and Haghi, A. K.; Eds. Education for a Digital World: Current Challenges and Future Possibilities. New York: Apple Academic Publishing; **2012**.

Luppicini, R.; and Myhill, C.; Student 2.0: a look at student values in a digital age. In: Education for a Digital World: Present Realities and Future Possibilities (133–145). Eds. Luppicini, R.; and Haghi, A. K.; New York: Apple Academic Publishing; **2012**.

Luppicini, R.; and Barber, P.; Inside student 2.0: student perspectives on navigating online and offline identities. In: Education for a Digital World: Present Realities and Future Possibilities (161–184). Eds. Luppicini, R.; and Haghi, A. K.; New York: Apple Academic Publishing; **2012**.

Luppicini, R.; and Lin, S.; Student 2.0 revisited: student perspectives on anonymity and privacy in the digital world. In: Education for a Digital World: Present Realities and Future Possibilities (185–196). Eds. Luppicini, R.; and Haghi, A. K.; New York: Apple Academic Publishing; **2012**.

McLuhan, M.; The Gutenberg Galaxy. Toronto: McGraw Hill; **1962**.

McLuhan, M.; and Powers, B.; The Global Village: Transformations in World Life and Media in the Twenty-First Century. Oxford: Oxford University Press; **1989**.

Nichols, R. G.; Toward a conscience: negative aspect of educational technology. *J. Visual/Verbal Languaging*. **1987,** *7*(1), 121–137.

Nichols, R. G.; Searching for moral guidance about educational technology. *Educ. Technol.* **1994,** *34*(2), 40–48.

Ratemyprofessors. Retreived August 1, 2013 from, **2013**.

Svensson, G.; and Wood, G.; Are university students really customers? When illusion may lead to delusion for all! *Int. J. Educ. Manage.* **2007,** *21(1),* 17–28

Trice, H.; and Beyer, J.; The Cultures of Work Organizations. New Jersey: Prentice-Hall; **1993**.

University of Ottawa. Retrieved August 1, 2013 from, http://www.uottawa.ca/all-about/canadasuniversity/, **2013**.

Whitehead, A. N.; The Aims of Education and Other Essays. New York: New American Library; **1949**.

Wolff, R. P.; The Ideal of the University. New York: Transaction Publishers; **1992**.

CHAPTER 3

BALANCING THE RIGHTS OF STUDENTS WITH DISABILITIES AND ACADEMIC INTEGRITY: CASE STUDIES FOR TEACHER EDUCATION FACULTY

MAUREEN E. SQUIRES

CONTENTS

3.1 INTRODUCTION

More people with disabilities are pursing postsecondary education than ever before (Baldwin, 2007; Papalia-Berardi, Hughes, and Papalia, 2002). These students major in myriad programs, including teacher education. Federal civil rights legislation paved the way for greater access to higher education. Section 504 of the Rehabilitation Act of 1973 prohibits discrimination "solely on the basis of disability," and the Americans with Disabilities Act (ADA) aims at creating a "fair and level playing field" for eligible people with disabilities (Hardman, Drew, and Egan, 2008). Faculty members in teacher preparation have a legal and ethical obligation to comply with Section 504 and ADA. Simultaneously, there is a legal and ethical imperative to prepare teacher candidates for the demands they will be expected to perform autonomously in the classroom. Both the No Child Left Behind Act of 2010 (NCLB and the Individuals with Disabilities Education Improvement Act of 2004 (IDEIA require local education agencies to employ highly qualified teachers. In turn, these mandates affect teacher preparation programs, which are responsible for preparing competent teacher candidates. Herein lies a challenge: Balancing the rights of teacher candidates with disabilities with professional standards that presumably lead to effective teaching. The purpose of this chapter is to prompt dialogue about this challenge and examination of academic programs. Legal, theoretical, and empirical sources will be presented to ground analysis of three case studies, with implications for one's own teaching.

3.2 STUDENTS WITH DISABILITIES AND HIGHER EDUCATION

Colleges and universities are made up of a diverse population. The number of students with disabilities in higher education has steadily increased over the past several decades (Brulle, Leyser, Vogel, and Wyland, 1998; Csoli and Gallagher, 2012; Leyser and Greenberger, 2008; Leyser, Greenberger, Sharoni, and Vogel, 2011). Some studies (Henderson, 2001; Vogel, Leyser, Burgstahler, Sligar, and Zecker, 2006) show students with learning disabilities (LDS) as the fastest growing subgroup, whereas others (Show and Scott, 2003) indicate students with attention deficit hyperactivity disorder (ADHD) and emotional disorders are increasing even faster.

Raue and Lewis' (2011) study contributes new data on a national level. They surveyed 2-year and 4-year degree-granting postsecondary institutions regarding students who identified themselves as a having a disability. They found 88 per cent of participating institutions enrolled students with disabilities. By disability category, 86 per cent of institutions enrolled students with specific LDs; 79 per cent, students with attention deficit disorder (ADD) or ADHD; 76 per cent with mobility limitations or orthopedic impairments; and 76 per cent with psychological conditions (Raue and Lewis, 2011). The overwhelming majority of higher education institutions in the study provided services for students under ADA. The distribution of disabilities in post-secondary education parallels that of P–12 schools. The most frequently reported disability was LD (31%), followed by ADD/ADHD (18%), psychological conditions (15%), and other health impairments (11%). This indicates that students with multiple and diverse needs are attending myriad institutions across the country. These findings support previous studies in the field.

An increasing proportion of these students with disabilities are entering teacher preparation programs, intending to become public P–12 educators (Baldwin, 2007; Csoli and Gallagher, 2012; Papalia-Berardi et al., 2002). Like their nondisabled peers, students with disabilities have various degrees of success in their academic programs. Regardless of classification, all teacher candidates must demonstrate competency and fulfill academic/professional requirements. According to the National Commission on Teaching and America's Future, "What teachers know and do is the most important influence on what students learn. Competent and caring teaching should be a student right" (as cited in Brulle, 2006, p. 3). The drive to staff all classrooms with "highly qualified teachers" was accompanied by enhanced professional standards and more rigorous certification exams to meet the demand for well-prepared teachers. (It is beyond the scope of this chapter to discuss whether teachers were truly deficient, thereby requiring such reform.) Areas under scrutiny include knowledge of subject matter and content, instructional skills and strategies, dispositions toward students (Brulle, 2006; Leyser and Greenberger, 2008); and effective communication (Gilbert, 1998; Wertheim, Vogel, and Brulle, 1998)—all of which are supported by National Council for Accreditation of Teacher Education and Teacher Education Accreditation Council.

Teacher education faculty are faced with two tasks: fulfilling legal obligations that protect the rights of students with disabilities and ensuring academic and professional integrity in the classroom and clinical setting. Such ethical issues are not always easy to balance. As more students with disabilities enter teacher preparation programs, it becomes increasingly important for faculty to understand their obligations and design appropriate programs. The following section provides an overview of relevant legal mandates.

3.3 ENSURING THE RIGHTS OF PERSONS WITH DISABILITIES

3.3.1 LEGISLATION

Section 504 of the Vocational Rehabilitation Act (1973) prohibits any institution or organization receiving federal funds from discriminating against persons on the basis of a disability. It is considered the first civil rights law for people with disabilities. Section 504 states, "No otherwise qualified person with a disability … shall, solely on the basis of disability, be denied access to, or the benefits of, or be subjected to discrimination under any program or activity provided by, an entity/institution that receives federal financial assistance" (P.L. 93–112). A qualified person is one who has a physical or mental impairment, has a record of such impairment, or is regarded as having such an impairment that substantially limits one or more major life activities. Major life activities include walking, seeing, hearing, speaking, breathing, learning, and working. This law provides equal access to federally funded programs for qualified persons.

Subpart E of Section 504 pertains to postsecondary education. It stipulates that qualified persons cannot be discriminated in recruitment, admission, and treatment. Specifically, students with disabilities must be offered housing, financial, and employment assistance, and non-academic services comparable to their nondisabled peers. Moreover, they must be provided academic adjustments; yet they still must meet academic requirements that are "essential to the instruction being pursued … or to any directly related licensing requirement." That means institutions are not obligated to fundamentally alter programs (e.g. change requirements or essential elements). While it is the responsibility of higher education institutions to uphold Section 504, they cannot do so without information from the individual with a disability. Students must self-identify or disclose their disability to the appropriate institutional representative. Specifically, they must provide documentation of their disability, the need for academic adjustments, and a list of requested modifications, auxiliary aids, or services.

The Americans with Disabilities Act (ADA) of 1990, modeled after Section 504, extends the rights of persons with disabilities to the private sector. It is intended to create a "fair and level playing field" for eligible, otherwise qualified persons with disabilities (Hardman et al., 2008). Eligible persons are those who have a physical or mental impairment that substantially limits a major life activity. Otherwise, qualified refers to persons who can "perform the most fundamental duties of the position (held or desired) with or without reasonable accommodation" (Gilbert, 1998, p. 323), that is, individuals who can meet the essential intellectual and performance standards of a given profession.

In regard to academics, ADA mandates postsecondary institutions to provide reasonable accommodations. Such adjustments "permit students with disabilities the opportunity to learn by removing barriers that do not compromise academic standards" (Ferris, n.d.). Reasonable accommodations are determined on a case-by-case basis. They can include modifying existing facilities; changing the schedule or location of classes; providing qualified readers; or adjusting examinations, to name a few (Gilbert, 1998). Accommodations are determined to be reasonable based on their effectiveness, that is, whether they meet academic needs of the individual. Teacher preparation programs are legally obligated to follow such nondiscriminatory laws.

3.3.2 LITIGATION

Numerous court cases in clinical fields (e.g. in teacher education or related fields) shed light on the application of disability legislation. In *Southeastern Community College v. Davis* (1979), a nursing candidate with a significant hearing impairment claimed the college was discriminatory on the basis of disability by rejecting her application for admission. The college claimed her hearing impairment "made it impossible for her safely to participate in a normal clinical training program or properly to care for patients" (Yanok, 1987, p. 49). The court found in favor of the college, ruling that the plaintiff was not "otherwise qualified" and stating that Section 504 does not prohibit an educational institution from requiring "reasonable physical qualification for admission to a clinical training program"

(Brulle, 2006, p. 4). This case illustrates the college's right to establish and maintain essential qualifications for admission to a clinical training program. As teacher preparation programs are necessarily clinical, *Southeastern Community College v. Davis* (1979) has implications for the field of education.

Similarly, other court case decisions have upheld institutions' right to set requirements for program completion or professional certification/licensure. In *Board of Curators, University of Missouri v. Horowitz* (1977), the court ruled a medical student could be dismissed from the academic program due to academic deficiencies. The university had afforded the student procedural due process; its decision to dismiss the student was neither arbitrary nor capricious. As such, the institution acted in accordance with the constitution. In *Pandazides v. Virginia Board of Education* (1992), a probationary teacher with a not-previously documented LD claimed the School Board was discriminatory on the basis of disability by denying her a license after she failed the required competency exam. The court found in favor of the Board of Education. It stated that the plaintiff had not demonstrated she was "otherwise qualified." Even with accommodations on the competency exam, she was unable to earn the requisite cut score. The plaintiff had not demonstrated that she could perform "essential functions" (in this case, written language functions) of a public school teacher as established by the State of Virginia. This case demonstrates that candidates for licensure cannot be "excused from meeting the legitimate and valid requirements for a teacher's license" (Leagle, 2012). Likewise, in the *United States v. South Carolina* (1977), the court upheld the constitutionality of using competency exams for the purpose of certification. The jury concluded that, specifically, the National Teacher Examination "did not have a discriminatory intent" (Yanok, 1987, p. 50). Courts have upheld institutions' rights to enforce academic standards.

Not only have courts ruled on academic requirements but also dispositions. In *Lipsett v. University of Puerto Rico* (1986), a medical student sued the university after being dismissed from the surgical program due to her abrasive manner with patients and fellow employees. The court found for the university, stating thus:

> There may be situations where academic fitness requires evaluation of subjective character traits for these may be valid components of the characteristics of a certain trade or profession and their evaluation

necessary to determine "academic" fitness for that type of profession. (Brulle, 2006, p. 5)

Education, as a human service, requires positive interactions and relationships with human beings. Many teacher preparation programs, accrediting bodies, and professional organizations specify and assess dispositions. This case points to the right of an institution to establish dispositional requirements of students. By extension, *Lipsett v. University of Puerto Rico* (1986) could have implications for some individuals with Asperger's Syndrome or emotional disorders where problems with social skills and communication difficulties are common.

Yet not all rulings favor postsecondary institutions. In *Bartlett v. New York State Board of Law Examiners* (1998), the plaintiff, who had an LD, sued for discrimination when her application for accommodations on the bar exam was rejected. The judge found in favor of the plaintiff, stating that the plaintiff's experts "have convinced me that the extra time provided to learning disabled applicants merely levels the playing field and allows these individuals to be tested on their knowledge; it does not provide them with an unfair advantage" (WrightsLaw, 2013). The judge went on to explain that the ADA and Section 504 do not guarantee an otherwise qualified person conditions that will allow him/her to pass the bar examination; rather, they guarantee a person the right to take the exam on a level playing field and pass it on his/her own. In teacher education, this means that faculty must provide reasonable accommodations to candidates with disabilities, yet they do not have to guarantee successful completion of the program. Teacher candidates with disabilities must meet program and certification requirements.

Similarly, in a previous case, *Gurmankin v. Costanzo* (1977), the court found "blindness in itself was not sufficient to disqualify a person from receiving a teaching certificate" (Wertheim et al., 1998, p. 295). The judge ruled that the plaintiff must be given the opportunity to take the qualifying licensure exam, thereby providing evidence of competency, just like her nondisabled peers. The court stated that deprivation of the right to take the competency exam had been arbitrary and violated due process (FindACase, 2013). Moreover, the judge ruled that the plaintiff be offered a position in the school district, which formerly held medical and personnel policies that excluded people who are blind from teaching. *Bartlett v. New York State Board of Law Examiners* (1998) and *Gurmankin v. Costanzo* (1977) underscore the constitutionality of access and reasonable accommodations. In addition, institutions of higher education must have procedures in place for reviewing program requirements, determining whether requirements are essential to the academic program, and providing reasonable accommodation (*Guckenberger v. Boston University* (1997)).

3.4 EXTANT LITERATURE

3.4.1 FACULTY PERSPECTIVES AND WILLINGNESS

Student success depends on myriad internal and external factors. For post-secondary students with disabilities, this includes faculty knowledge of, disposition toward, and willingness to make accommodations for students with disabilities (Leyser and Greenberger, 2008; Leyser et al., 2011). One body of literature indicates higher education faculties have minimal knowledge of disability laws (Leyser, Vogel, Brulle, and Wyland, 1998; Leyser, Vogel, Wyland, Brulle, Sharoni, and Vogel, 2003; Vogel et al., 2006). Other studies have reported both nonsupportive and supportive faculty attitudes toward students with disabilities. Some reports showed faculty were "especially skeptical and mistrusting of students with non-visible disabilities," including LDs, ADHD, and emotional disorders (Leyser et al., 2011, p. 163). Yet, for the most part, faculty held positive attitudes and were willing to provide reasonable accommodations. Not all accommodations are viewed equally. Faculty were more likely to provide teaching and examination accommodations and less willing to offer exclusive extra credit assignments, provide copies of lecture notes, overlook poor spelling, punctuation, or grammar, or make substitutions for required courses (Leyser et al., 2003; Sweener, Kundert, May, and Quinn, 2002). This suggests that faculty are willing to provide accommodations, accommodations that are perceived as not lowering program standards.

Studies in teacher education yield similar findings. Baldwin's (2007) research revealed that faculty generally found classroom accommodations for teacher candidates acceptable; however, they considered field accommodations ineffective and unethical. Faculty rated accommodations effective if they helped teacher candidates be successful and maintained the integrity of the program. Such accommodations could be categorized as adapting instruction (or the presentation of concepts) and assistance on preparation of assignments (e.g. written and oral directions, outlines or notes for lectures, submission of drafts, extended time on exams, and exams in alternate locations) (Baldwin, 2007). Ineffective accommodations were those that modified the program (e.g. adapted grading, extra credit to boost course grades, an alternative requirement for student teaching, and substitution of a course for a required course), thereby lowering expectations for students with disabilities (Baldwin, 2007). Faculty perceived accommodations unethical if they were not transferrable to field-based settings, that is, if accommodations used to complete coursework were not available on the job when teacher candidates would be responsible for independent teaching (Baldwin, 2007). Other acceptable accommodations for teacher candidates include assistive technology and alternative demonstrations of competency; ignoring GPA requirements was not supported by higher education faculty (Brulle et al., 1998).

In another body of literature, researchers explored factors that influenced faculty willingness to provide accommodations. Faculty knowledge (Bigaj, Shaw, and McGuire as cited in Leyser and Greenberger, 2008), the number of students requesting accommodations (Bourke, Strenhorn, and Silver as cited in Leyser and Greenberger, 2008), and several demographic variables affect willingness. As synthesized by Leyser and Greenberger (2008), gender, teaching experience, training, academic rank, and academic discipline influence faculty willingness to make accommodations. Data suggest women; faculty with more experience teaching students with disabilities; faculty with more professional development on disabilities; faculty who do not hold doctorates; and faculty in colleges of education are more willing to provide accommodations than their counterparts. Further, students in teacher preparation programs are working not only with faculty but also with supervisors and cooperating teachers who often have mixed reactions about teachers with disabilities.

3.5 PREVALENT ASSUMPTIONS

3.5.1 *THE STIGMA OF HAVING A DISABILITY*

Disclosure is a double-edged sword. The only way postsecondary students can receive accommodations or auxiliary aids is to self-identify; yet, identifying oneself as having a disability can be stigmatizing. For many years, society held that individuals with disabilities were not "normal." At different times in history, people with disabilities were considered possessed by demons, sick, deviant, suffering, infantile, ugly, pitiable, and needing to be cured (Connor and Bejoian, 2006; Hardman et al., 2008; Valle, and Connor, 2011; Wolfensberger, 1975). While society's position on disability has greatly changed over the past few decades, such negative perceptions of disability are still deeply ingrained in US culture. For example, students who are labeled with an LD often experience lower teacher expectations. Their teachers are likely to "provide more rewards and fewer punishments, exhibit more pity and less anger toward that student, and have increased expectations of failure rather than success" (Clark, 1997 in Ferri, Keefe, and Gregg, 2001, p. 27). For persons with disabilities in the role of teacher, this stigma persists. Students evaluating the quality of three teachers ranked the teacher who disclosed an LD less favorably than the teacher with a physical disability or no disability (Reilly, Wennet, Murphy, Thierauf, Gaugler, and Godfrey, 1998). Teacher candidates with disabilities are aware of these stereotypes.

One line of research, focused on self-disclosure of persons with LDs, reveals that the decision to disclose disability is highly personal and carefully deliberated. Choosing when and to whom to disclose this information is often complicated. Some teacher candidates felt safe disclosing to their academic instructors and did so with "minimal concerns or reservations" (Csoli and Gallagher, 2012, p. 73). Other teachers and teacher candidates debated whom to tell: faculty, supervisors, cooperating teachers, administrators, colleagues, parents, and students (Ferri et al., 2001; Gilbert, 1998; Valle, Solis, Volpitta, and Connor, 2004; Vogel and Sharoni, 2010). Some individuals feared that disclosure would negatively affect them. For instance, they believed they would not be given a fair chance (Gilbert, 1998) and would be viewed as dumb or stupid (Ferri et al., 2001). Other persons were concerned that they would be considered lazy or not taken as a professional or authority figure (Valle et al., 2004). These persons perceived disclosure as costly and often chose (or were forced by society) to remain "closeted." Yet, other teacher candidates elected not to disclose to prove a point; they wanted to dispel stereotypes and show they could be successful without accommodations (Valle et al., 2004).

3.6 PERCEIVED DEFICITS AND UNDERLYING ASSUMPTIONS

For many years, a deficit-based model has prevailed in the field of disabilities. This paradigm posits disability as a defect within the individual that needs to be corrected or overcome for the individual to be successful (Ferri, Connor, Solis, Valle, and Volpitta, 2005; Pfeiffer, 2002; Valle et al., 2004). This model presents persons with disabilities as abnormal and positions them as victims of their disability. As such, they somehow need to compensate for what they lack. A deficit-based model is limited in many ways, primarily because it emphasizes perceived weaknesses and ignores the influence of sociocultural contexts (through which we construct meanings of disability). In opposition to such a reductionist model, other theories have been developed to study disability. One alternative is the strengths-based model. This model focuses on what individuals with disabilities can do and aims to build upon the capacities they have or can develop (Lietz, 2004). (Other frameworks include the medical, charity, legitimacy, or economic model, to name a few. It is beyond the purview of this chapter to discuss all models of disability.)

In addition to the paradigm in which disabilities are perceived as weaknesses, underlying assumptions about good teachers can pose challenges to teacher candidates with disabilities. The ability of teachers with disabilities has been questioned based on presumptive knowledge of the disability. At one point, perhaps still, it was assumed that a teacher with visual or hearing impairments could not manage classroom behavior (Papalia-Berardi et al., 2002); a teacher with an emotional disorder was a risk to students (McGee and Kauffmann, 1989); and a teacher with an LD lacked basic literacy, mathematical, or cognitive abilities necessary to instruct students in these areas (Gilbert, 1998; Papalia-Berardi et al., 2002; Riddick, 2003; Wertheim et al., 1998). Yet, such definitive statements, gross overgeneralizations, cannot be made. Each case must be considered individually—each teacher candidate must be given an opportunity to demonstrate his/her capability. The mere presence of a disability cannot and should not disqualify a person from pursuing a teaching career.

As well, underlying assumptions about teacher preparation have resulted in barriers for individuals with disabilities who choose the teaching profession. For example, some professionals assume that scores on standardized exams are predictive of teaching effectiveness. Academic degree programs require minimum scores for admission and certification bodies set qualifying passing scores. By definition, standardized tests are administered and scored in a standard manner. That means, all students must demonstrate their knowledge in the same way, which disregards multiple ways of expressing knowledge. Norm-referenced standardized tests present an additional concern. The scores for these exams are "normed" on a particular group of people. This group is supposed to represent the pool of test-takers, but may not be representative of the actual population. Therefore, score interpretations can be skewed in favor of the normed group, which often includes typical, not diverse, students. Teacher candidates with LDs often struggle on such basic competency or certification exams. Critics of standardized tests argue that they more likely measure disability than ability and question whether standardized tests have construct or predictive validity (Otis-Wilborn, Cates, Proctor, and Kinnison, 1991; Papalia-Berardi et al., 2002; Riddick and English, 2006).

A second assumption implies a link between performance in the college classroom and clinical settings. Some believe that GPA (as determined by performance on course assignments) predicts teaching performance (Otis-Wilborn et al., 1991). This fallacy neglects to consider that course grades (particularly, when determined in traditional, undifferentiated ways) can underestimate a student's true knowledge or ability. GPA does not necessarily or accurately predict teaching practice. Teaching is a complex art and science. It is more than knowing and explaining theories of learning and development; it is interacting with human beings, critically assessing self and others, and implementing best practices for particular students.

Another assumption, two-fold, is that literacy competency precedes learning and that literacy is directly related to performance in college classroom and clinical settings (Otis-Wilborn et al., 1991). What follows is the argument that teacher candidates with below-standard literacy skills are not as capable of teaching reading and writing as their nondisabled colleagues. That is, teacher candidates with LDs, in particular, dyslexia, would be unable to assess or provide feedback on student written work (Riddick and English, 2006; Wertheim et al., 1998). A parallel argument: teacher candidates need exemplary literacy skills as they are expected to model these for children (Riddick, 2006). While it may seem common sense to link the two, evidence does not illustrate a connection between the literacy performance of teachers and the literacy standards of the children they teach (Riddick and English, 2006). On the contrary, researchers (Duquette, 2000; Morgan and Rooney, 1997) found that teacher candidates with dyslexia knew their strengths and weaknesses, identified and used appropriate compensatory strategies, did not rely on special assistance, and held high literacy expectations for children.

By definition, individuals with learning LDs are of average or above-average intelligence with a "disorder in one of more of the basic psychological processes involved in understanding or in using language, spoken or written, that may manifest itself in the imperfect ability to listen, think, speak, read, write, spell, or do mathematical calculations" (IDEIA, 2004). Persons with LDs are not "dumb" or uneducable; their brains simply process information differently, which does not make them incapable of teaching others. Adding a new voice to this discussion, teachers with disabilities consider themselves capable and effective in the classroom. In numerous studies (Ferri et al., 2001; Leyser and Greenberger, 2008; Vogel and Sharoni, 2010), teachers with LDs believed themselves successful. They viewed their disability as an asset, making them better teachers because they could empathize with students with disabilities. They also used their personal experience (e.g. modeling effective compensatory strategies) to help students achieve success.

This supports previous research on satisfactorily teaching. Karp and Keller (1998) interviewed 25 educators, male and female, with various disabilities, at different career stages, practicing at various educational positions (e.g. childhood educator, adolescence educator, special educator, counselor, or professor). They found that nearly all of the teachers "could satisfactorily perform the professional responsibilities of their professions" (p. 84). (Satisfactoriness was determined by length of service; commendations; or evaluations by supervisors, promotions, and personal stories.)

Although the formerly discussed deficit-based model and underlying assumptions exist, currently, they do not dominate higher education. Research (Leyser and Greenberger, 2008; Leyser et al., 2011; Vogel and Sharoni, 2010) indicates that the overwhelming majority of faculty members believe teachers with disabilities can be just as effective in the field as teachers without disabilities. These barriers are presented to prompt readers' critical self-reflection. Since thought and action are intertwined, it is important for teacher educators to examine their own beliefs and consider how they affect the learning experience of their students.

3.7 THROUGH A CRITICAL LENS

At the center of critical theory are two central beliefs. First, the notion that reality is socially constructed, situated in multiple interrelated contexts, and not objective. Second, the idea that hegemony perpetuates privilege for some people and oppression of others. Critical theorists believe that society must be problematized and reformed; otherwise, culture will continue to reproduce inequality. Critical theory is multifaceted, serving both as a way to analyze phenomena and a vehicle to advance social change. Critical theorists expose and challenge social, historical, political, and economic conditions of oppression. They critique unequal power relations and underlying assumptions and suggest ways to reform society so that individuals can be emancipated from oppression (Bogdan and Biklen, 1998; deMarrais and LeCompte, 1999; Gall, Gall, and Borg, 2005; Lichtman, 2006).

The goal of teacher education is to prepare future educators who are caring and competent. Regardless of ability, all teacher candidates should be held to the same high standard. Any less would be a disservice to P–12 students. Yet higher education faculty should not automatically accept standards as markers of effective teaching. Educators should carefully consider who sets the standards, how they are assessed, and how good teaching is defined. Such critical analysis (often used in race and gender studies) has recently entered the study of disability in education. Further, it should be noted that focus on standards brings attention to just that, the end product. It does not take into account the process (e.g. qualities such as intellectual curiosity, persistence, and time management) by which the standards were achieved. Arguably, such nonacademic factors affect success as a student and teacher. Overemphasis on standards (Riddick and English, 2006) and ability testing and labeling (Ferri and Connor, 2005) can be used as exclusionary practices to keep marginalized people out of the field. In this way, persons who determine entrance criteria (those already "in") are able to maintain their sense of normalcy and authority.

To a degree, even the inclusion movement in education, considered by many people as a measure to reverse the marginalization of students with disabilities, maintains the imbalance of power and privilege. Asante (n.d.), a community activist and organizer argued,

> It is generally accepted that "Inclusion" means inviting those who have been historically locked out to "come in" ... A weakness of this defini-

> tion is evident. Who has the authority or right to "invite" others in? And how did the "inviters" get in? Finally, who is doing the excluding? It is time we both recognize and accept that we are all born "in"! No one has the right to invite others in! It definitely becomes our responsibility as a society to remove all barriers which uphold exclusion since none of us have the authority to "invite" others "in"!

Similar to critical race scholars' understanding of "race," disability studies scholars maintain that "ability" is socially, not biologically, constructed. Disability is a "discursive construction" (Thomson as cited in Ferri et al, 2005, p. 63), and for too long the disability discourse has centered on pathological or medical models. Such paradigms serve to maintain the status quo by perpetuating oppressive stereotypes and "keeping out" persons with disabilities. Simultaneously, individual and cultural behaviors, capital, and needs are ignored because they are outside, and threaten, the "norm" (Ferri, 2008; Ferri and Connor, 2005; Ferri et al., 2005; Zion and Blanchett, 2011). Disability studies scholars propose ways of "re-theorizing and re-imagining dis/ability and normalcy ... that can serve an "irritant to the dominant paradigm of knowledge" about disability (Ferri, 2008, p. 498–499).

3.8 SUGGESTIONS FOR TEACHER EDUCATION

A review of the literature reveals several ways of improving both academic programs and the experience of students with disabilities in teacher preparation programs. As noted earlier, dis/ability is a social construct. Its meaning is neither fixed nor value-free; rather, it is given meaning by people in a given time and place. To revise the typical negative connotation of disability, people must acknowledge and change underlying assumptions. This requires a willingness to examine previous lived experiences with persons with disabilities, attitudes, and biases. It is essential that teacher education faculty critically appraise their understandings of dis/ability and reflect on the connection between their beliefs and classroom practices (Csoli and Gallagher, 2012; Ferri et al., 2001; Ferri et al., 2005). Such critical reflection can be used to enhance personal and program development.

Professional development can also be provided for teacher education faculty. Research indicates some faculties have few experiences teaching students with disabilities and limited knowledge of support services (Baldwin, 2007; Leyser et al., 1998; Leyser et al., 2003; Vogel et al., 2006). Studies also suggest that training positively affects attitudes and classroom practice (Leyser and Greenberger, 2008). Since faculty have different needs, training should be tailored to fit their preferences. Faculty can engage in professional development in multiple ways: large group seminars, small group workshops, individual study; face-to-face meetings, online courses, training videos, etc. When designing professional development, plan follow-up sessions and ways of helping faculty transfer learning from the training to, and sustain it in, the classroom.

Teacher preparation programs need to examine, clarify, and communicate standards and essential elements to all teacher candidates. Upon entering a program, teacher candidates should know what they are required to do and at what level. This includes performance and academic standards, in addition to standards for certification (Baldwin, 2007; Brulle, 2006; Brulle et al., 1998; Gilbert, 1998; Papalia-Berardi et al., 2002). Such essential standards and elements may consist of grade point average, requisite scores on standardized competency or licensure exams, capstone experiences, fieldwork requirements, and dispositions. These minimal competencies should be well defined and widely published. Appeals or dismissal policies and procedures also should be clear, unbiased, and widely disseminated. Institutions of higher education are legally and ethically obligated to follow due process. Further, it is recommended that programs maintain extensive records of data for decisions, accommodations provided, communication exchanged, and notes from meetings (Brulle, 2006; Wertheim et al., 1998).

Additionally, programs should consider and design reasonable accommodations. Reasonable accommodations can be used in the college classroom and the field, as determined by the program based on the individual's needs and in accordance with legislation. General accommodations for persons with disabilities include preferential seating, enhancement of auditory or visual instruction, extended time for exams or assignments, tutoring, and increased opportunities for practice (Otis-Wilborn et al., 1991). In clinical settings, consideration could be made for matching teacher candidates with appropriate sites, classroom demands, and supportive cooperating teachers (Wertheim et al., 1998). Teacher candidates with disabilities have the right to reasonable accommodations. Faculty members have the responsibility to provide such accommodations, accommodations that simultaneously give students with disabilities a level playing field and prepare them to perform the essential functions of the position.

Finally, provide teacher candidates training in self-regulation and self-advocacy. To be successful in higher education, students need to have more than academic skills. College success also is influenced by nonintellectual factors such as time management, work habits, and planning (Foley, 2006). For students with disabilities, nonacademic skills include describing the disability, explaining how the disability affects learning, accessing services, suggesting effective accommodations, and discussing academic concerns with professors (Foley, 2006; Papalia-Berardi et al., 2002). Students must know their rights, understand their disability, and effectively communicate to receive accommodations at postsecondary institutions.

3.9 CASE STUDIES IN HIGHER EDUCATION

The following vignettes represent students with documented and alleged disabilities in undergraduate or graduate teacher preparation programs. I have written case studies to capture students' experiences throughout programs. Case studies represent a variety of students: male and female; different subject areas; diverse prior experiences with special education; various degrees of self-advocacy; and different identified or presumed disabilities. These case studies can be used to explore personal beliefs and program elements in light of the theories and legislation presented.

3.9.1 CASE STUDY 1: JOELLE

Joelle was accepted into a Master of Science for Teachers program in adolescence education. Acceptance, as is standard for the program, was based on team review of an application packet (including a form, written statement of purpose, current resume, three letters of recommendation, official transcripts, and scores from the GRE, MAT, or GMAT) and an on-campus interview. Graduate admissions require the following: a baccalaureate degree from an accredited institution, a GPA of at least 2.5 on a 4.0 scale, satisfactory scores on admission tests, and satisfactory application packet and interview. This is a three-semester program, the first two semesters are devoted to coursework and fieldwork, and the last is student teaching.

In the first semester, several professors spoke with Joelle about her writing. Her papers were disorganized and contained grammar and spelling mistakes. Her written work did not match her oral ability or the depth of her knowledge. In one-on-one conversations and presentations to the class, Joelle could present ideas coherently. Joelle shared that she used to have an LD in written expression. She had received special education services in middle school and opted out of services in high school. She could identify her strengths and weaknesses and had been relying on the computer's spell check and grammar check to correct writing errors. Joelle was self-aware, to a degree, and willing to disclose her disability, although this was not on record at the college.

In the second semester, Joelle continued to have trouble with writing. Some professors granted her extensions on papers, offered feedback and strategies, and recommended she visit Student Support Services. Joelle's precarious GPA put her in jeopardy of not student teaching the following semester. (Graduate students must maintain a "B" or better in all courses to continue in the program.) Concerned about her academic progress and overall professional development, Joelle requested a current cognitive evaluation from the Student Support Services office. She received the results and shared them with some professors. The report indicated that she had a learning disability. Specifically, she had difficulty with written expression as evidenced by grammar, punctuation, and spelling errors and poor paragraph organization. In addition, she had difficulty with short-term memory. She believed the evaluation was valid and reliable as it supported prior testing and her current experience.

Earning grades of "B" or better in classes often required conferences with instructors, proofreading from peers, and re-writing and re-submitting assignments. Joelle did not earn a "B" or better in each course, primarily due to low marks on written work, not lack of content knowledge. Consequently, she could not proceed to student teaching until she raised her GPA. Through meetings with the student and several faculty members, it was decided that Joelle not retake the course. Rather, she would complete an alternate, co-designed requirement. We all agreed that coherent writing is essential for teachers. Accordingly, Joelle opted to research relevant compensatory strategies and take a composition course to learn more about writing. She learned how to use software such as Dragon NaturallySpeaking to enhance writing and monthly organizers to facilitate memory and manage time. She successfully completed the alternate requirement, student teaching (albeit a semester behind her cohort), and the program.

This case surfaces several questions and concerns. What are the essential functions of a high school teacher? What language and memory demands are placed on classroom teachers? Although teachers can use spellcheck, colleagues as proofreading tools, or detailed notes to plan classes, how will teachers function in-the-moment when writing notes on the board or holding multiple pieces of information in short-term memory? How will teachers help students enhance written communication if they cannot identify their own writing errors? Should a graduate student be asked to complete a course in an identified weakness to progress in the program? Should teacher candidates be allowed to rewrite (based on instructor feedback and conferences) and resubmit papers to improve their grades when they will not have this opportunity in the field? A misspelled email to an administrator or letter with incomplete sentences sent home to a parent cannot be retracted.

3.9.2 CASE STUDY 2: AUDREY

Like Joelle, and under the same conditions, Audrey was accepted into a Master of Science for Teachers program in adolescence education. Several faculty members voiced concerns about Audrey during the interview process, noting she had problems with social interactions. Upon entering the interview room, Audrey did not smile or make eye contact, rather stared straight ahead, as if peering through the faculty members. She neither introduced herself nor initiated a handshake. Her facial expression and modulation remained flat throughout the conversation. The interview team was concerned yet hesitant to reject her application based on a single encounter, choosing instead to base its decision on her entire application packet and presuming she was nervous at the interview.

Hoping to build connections with each new teacher candidate, faculty arrived at orientation intending to learn more about the individuals. When asked to share about her summer, Audrey spoke despairingly about migrant workers and persons of color. Her remarks were so off-putting that both faculty and fellow teacher candidates were taken aback; yet, Audrey did not sense the inappropriateness of her comments or the tension in the room. Faculty members called for a meeting with Audrey to discuss the event. She did not recall making the remarks or see them as a problem. Later in the meeting, Audrey pulled her shirt up, revealing her bare stomach, to wipe tears from her eyes. As during orientation, it appeared that Audrey had limited social awareness.

Audrey was a bright student and was strong in her subject area. She was passionate about her content area. She shared no former or current documentation of a disability with faculty. Professors grew more concerned as the semester progressed. Audrey rarely initiated conversations with peers or instructors; chose to sit on the periphery of the classroom; did not interact with the audience when presenting micro lessons; and disregarded greetings in the hallway. In a private conversation with Audrey, one professor discussed the importance of making eye contact and connections with people, especially adolescents; yet, during that meeting Audrey did not make eye contact with the professor.

Audrey did not earn the required "B" in all of her courses necessary to progress in the program. Failing grades were posted because she had not submitted assignments, thereby not meeting all course requirements. This meant she had to repeat classes before student teaching, which would extend the length of her program. At this point, Audrey is considering withdrawing from the program.

Effective teachers connect with students. They are approachable and take interest in adolescents. Admittedly, there is no single way for teachers and students to bond. What does connecting with students look like and sound like in a middle or high school classroom? Given that content knowledge is not enough for effective teaching, what essential dispositions should teacher candidates have so they are successful with all students? Should these dispositions take into account individuals' socioeconomic and cultural backgrounds? How can P–12 schools and postsecondary institutions be inclusive of individuals with diverse social skills? What can faculty do to mentor teacher candidates, helping them develop and demonstrate an ethic of care? How can faculty monitor pre-service teachers' dispositions, and what protocols should be in place to support and dismiss students from teacher preparation programs?

3.9.3 CASE STUDY 3: COREY

Corey was accepted into a combined bachelor of arts/master of science in teaching program in adolescence education. Acceptance was based on a standard application packet consisting of a common form, official high school transcripts, SAT/ACT scores, a letter of recommendation, and a writing sample. Undergraduate admissions requires the following: an 85 high school average and combined (critical reading and math) SAT scores of 1050. Students in this program must maintain a 2.75 GPA as an undergraduate and a 3.0 as a graduate student.

Corey's introduction to the education program was a foundations in education course. His professors noted a concern about his writing in the first week of the semester. His initial journal posting was disorganized and nearly incoherent, containing numerous grammar, spelling, punctuation, and capitalization errors. Professors provided feedback and suggestions for improvement, yet his second journal showed no growth. When professors spoke to Corey, they learned that he did not understand how to apply the feedback. He acknowledged he was a weak writer but was unable to describe his particular challenges or effective compensatory strategies. He said he had struggled his entire life and accepted his communication challenges as inevitable and incapacitating to a degree. Professors offered to meet with Corey to discuss feedback and writing strategies. They invited him to revise and resubmit work, an opportunity of which he did not take advantage. Several times, he missed appointments with the professors and neglected to reschedule. He earned a "C-" in the course. They also recommended he utilize Student Support Services, which he did not consult during the semester.

Professors in his content area noted Corey's writing challenge as well. They made similar suggestions and offers to Corey as education faculty. They also noticed his deep, complex understanding of the discipline. Content area faculty repeatedly commented that his conceptual understandings were innovative and dynamic. A year later, Corey recognized that he had ignored the help offered to him in the past and now wanted to avail himself of it. He took a grammar course to enhance his knowledge of writing. Believing he had a writing disability, Corey started to look into opportunities (tutoring and assessment) provided by Student Support Services. He even started researching model support programs for students with learning disabilities at other colleges. Currently, Corey is in jeopardy of beings dismissed from the program as he did not meet the GPA requirement.

Legally, institutions of higher education do not have to seek out and refer students with special needs. And, as previously discussed, disclosure of known disabilities is a responsibility of the college student. Ethically, what should faculty do to best support their students, particularly students who are suspected of having learning disabilities? How much and why type of individualized instructional support or remediation should faculty provide a student? What information and resources, upon entering and throughout the program, would help students be successful? What partnerships exist and can be forged between teacher preparation programs and Student Support Services to best meet the needs of students?

3.10 CONCLUSION

Teacher education programs have the weighty task of strengthening public education by preparing teacher candidates who are caring and competent. In this process, teacher education programs serve two audiences: the teacher candidates matriculated in degree programs and the future P–12 students whom they will teach. Due to, in part, legislation such as ADA and Section 504, more people have been guaranteed the right to equal access and level playing fields in higher education. Some of these people have entered teacher preparation programs. By law, faculty cannot discriminate against teacher candidates on the basis of disability. They must provide reasonable accommodations to otherwise qualified persons and ensure due process. While many faculty are willing to provide accommodations, other faculty question whether providing accommodation in college courses sets up teacher candidates for future failure (when they are the sole person responsible for their classroom).

Individuals with disabilities deserve the chance to prove their capability. They should not be presumed deficient, therefore unable to perform the essential elements of their chosen profession. Such deficit-based thinking disenfranchises persons with disabilities. At the same time, standards for admission and completion of teacher education programs should be rigorous for all people. P–12 students deserve to be taught by the best teachers. It is our responsibility, as faculty of teacher education programs, to balance our commitment to teacher candidates and P–12 students. This is a complex and dynamic responsibility, one that merits serious attention. It calls for a critical examination of both underlying beliefs and instructional practices, at the personal level, and requirements and policies of teacher preparation programs.

KEYWORDS

- **Disabilities**
- **Faculty members**
- **P–12 students**

REFERENCES

Asante, S.; (n.d.). What is inclusion? *The Inclusion Network.* Retrieved from http://www. inclusion.com /inclusion.html

Baldwin, J. L.; Teacher candidates with learning disabilities: effective and ethical accommodations. *Teach. Educ. Spec. Educ.* **2007,** *30(3),* 128–141.

Bogdan, R.; and Biklen, S.; Qualitative Research for Education: An Introduction to Theory and Methods. 3rd Ed. Boston, MA: Allyn & Bacon; **1998**.

Brulle, A. R.; Teacher candidates with disabilities: Guidance for teacher educators. *Act. Teach. Educ.* **2006,** *28(3),* 3–10.

Brulle, A. R.; Leyser, Y.; Vogel, S.; and Wyland, S.; Competencies and accommodations: faculty attitudes toward students with learning disabilities in teacher preparation. *Crit. Iss. Teach. Educ.* **1998,** *7,* 24–32.

Connor, D.; and Bejoian, L.; Pigs, pirates, and pills: using film to teach the social context of disability. *Teach. Except. Child.* **2006,** *39(2),* 52–60.

Csoli, K.; and Gallagher, T. L.; Accommodations in teacher education: perspectives of teacher candidates with learning disabilities and their faculty advisors. *Except. Educ. Int.* **2012,** *22(2),* 61–76.

deMarrais, K.; and LeCompte, M.; Theory and its influences on the purposes of schooling. In: The Way Schools Work: A Sociological Analysis of Education. 3rd Ed. USA: Allyn & Bacon; **1999,** 1–42 p.

Duquette, C.; Examining autobiographical influences on student teachers with disabilities. *Teach. Teach.: Theor. Pract.* **2000,** *6(2),* 215–228.

Ferri, B. A.; Changing the script: race and disability in Lynn Manning's weights. *Int. J. Inclu. Educ.* **2008,** *12(5–6),* 497–509.

Ferri, B. A.; and Connor, D. J.; Tools of exclusion: Race, disability, and (re)segregated education. *Teach. Coll. Rec.* **2005,** *107(3),* 453–474.

Ferri, B. A.; Connor, D. J.; Solis, S.; Valle, J.; and Volpitta, D.; Teachers with LD: ongoing negotiations with discourses of disability. *J. Learn. Disabil.* **2005,** *38(1),* 62–78.

Ferri, B. A.; Keefe, C. H.; and Gregg, N.; Teachers with learning disabilities: a view from both sides of the desk. *J. Learn. Disabil.* **2001,** *34(1),* 22–32.

Ferris State University. (n.d.) Reasonable Accommodations Defined. Retrieved from http://www. ferris.edu/HTMLS/colleges/university/disability/terms-concepts/reasonable accommodation.html

FindACase. Gurmankin v. Costanzo. Retrieved from http://de.findacase.com/research/wfrmDocViewer.aspx/xq/fac.19770225_0040182.C03.tm/qx, **2013**.

Foley, N. E.; Preparing for college: Improving the odds for students with learning disabilities. *Coll. Stud. J.* **2006,** *40(3),* 641–645.

Gall, J.; Gall, M.; and Borg, W.; Critical-theory research. In: Applying Educational Research. 5th Ed. Boston, MA: Allyn & Bacon; **2005,** 380–405 p.

Gilbert, S. L.; Another type of diversity: A student teacher with a learning disability. *Int. J. Qual. Stud. Educ.* **1998,** *11(2),* 323–340.

Hardman, M.; Drew, C.; and Egan, M.; Human Exceptionality: School, Community, and Family. 9th Ed. Needham Heights, MA: Allyn & Bacon; **2008**.

Henderson, C.; College Freshmen with Disabilities: A Biennial Statistical Profile. Retrieved from http://cms.hutchcc.edu/uploadedFiles/StudentServices/DisabilityServices/colfresh.pdf, **2001**.

Individuals with Disabilities Education Improvement Act [IDEIA] of 2004, 300 U.S.C. § A.300.8.c.10 (2004). Retrieved from http://idea.ed.gov/explore/view/p/%2Croot%2Cregs%2C300%2CA%2C300%252E8%2Cc%2C10%2C

Karp, J.; and Keller, C.; Preparation and employment experiences of educators with disabilities. In: Eds. Anderson, R.; Keller, C.; and Karp, J.; Enhancing Diversity: Educators with Disabilities. Washington, DC: Gallaudet University Press; **1998,** 74–88 p.

Leagle, Inc.; Pandazides v. Virginia Board of Education. Retrieved from http://www.leagle.com/decision-result/?xmldoc/19921598804FSupp794_11489.xml/ docbase/CSL-WAR2-1986-2006, **2012**.

Leyser, Y.; and Greenberger, L.; College students with disabilities in teacher education: faculty attitudes and practices. *Eur. J. Spec. Needs Educ.* **2008,** *23(3),* 237–251.

Leyser, Y.; Greenberger, L.; Sharoni, V.; and Vogel, G.; Students with disabilities in teacher education: changes in faculty attitudes toward accommodations over ten years. *Int. J. Spec. Educ.* **2011,** *26(1),* 162–174.

Leyser, Y.; Vogel, S.; Brulle, A.; and Wyland, S.; Faculty attitudes and practices regarding students with disabilities: two decades after implementation of Section 504. *J. Postsec. Educ. Disabil.* **1998,** *13(3),* 5–19.

Leyser, Y.; Vogel, S.; Wyland, S.; Brulle, A.; Sharoni, V.; and Vogel, G.; American and Israeli faculty attitudes and practices regarding students with disabilities: a cross-cultural study. In: Learning Disabilities in Higher Education and Beyond: International Perspectives. Eds. Vogel, S. A.; Vogel, G.; Sharoni, V.; and Dahan, O.; Baltimore, MD: York Press; **2003,** 201–225 p.

Lichtman, M.; Qualitative Research in Education: A User's Guide. Thousand Oaks, CA: Sage Publications; **2006**.

Lietz, C. A.; Resiliency based social learning: a strengths based approach to residential treatment. *Resident. Treat. Child. Youth.* **2004,** *22(2),* 21–36.

McGee, K. A.; and Kauffman, J. M.; Educating teachers with emotional disabilities: a balance of private and public interests. *Teach. Educ. Spec. Educ.* **1989,** *12(3),* 110–116.

Morgan, E.; and Rooney, M.; Can dyslexic students be trained as teachers? *Support Learn.* **1997,** *12,* 28–31.

Otis-Wilborn, A. K.; Cates, D.; Proctor, T.; and Kinnison, L.; Preparing students with disabilities as special educators. *Teach. Educ.* **1991,** *4(1),* 89–101.

Papalia-Berardi, A.; Hughes, C. A.; and Papalia, A. S.; Teacher education students with disabilities: participation and achievement factors. *The J. Teach. Educ. Div. Counc. Except. Child.* **2002,** *25(1),* 23–31.

Pfeiffer, D.; The philosophical foundations of disability studies. *Disabil. Stud. Q.* **2002,** *22(2),* 3–23.

Raue, K.; and Lewis, L.; Students with disabilities at degree-granting postsecondary institutions (NCES 2011–018). In: U.S. Department of Education, National Center for Education Statistics. Washington, DC: U.S. Government Printing Office.; **2011**.

Reilly, N. P.; Wennet, C.; Murphy, M.; Thierauf, M.; Gaugler, G.; and Godfrey, K. J.; Teaching evaluation of persons with disabilities differ according to the nature of disability. *J. Rehabil.* **1998,** *64(1),* 38–41. Retrieved from http://www.thefreelibrary.com/Teaching+evaluations+of+persons+with+disabilities+difer+according+to...-a020382087

Riddick, B.; Experiences of teachers and trainee teachers who are dyslexic. *Int. J. Inclus. Educ.* **2003,** *7(4),* 389–402.

Riddick, B.; and English, E.; Meeting the standards? Dyslexic students and the selection process for initial teacher training. *Eur. J. Teach. Educ.* **2006,** *29(2),* 203–222.

Show, S. R.; and Scott, S. S.; New directions in faculty development. *J. Postsec. Educ. Disabil.* **2003,** *17,* 3–9.

Sweener, K.; Kundert, D.; May, D.; and Quinn, K.; Comfort with accommodations at the community college level. *J. Dev. Educ.* **2002,** *25(3),* 12–14, 16, 18, 42.

Valle, J. W.; and Connor, D. J.; Rethinking Disability: A Disability Studies Approach to Inclusive Practices. New York: McGraw-Hill; **2011**.

Valle, J. W.; Solis, S.; Volpitta, D.; and Connor, D. J.; The disability closet: Teachers with learning disabilities evaluate the risks and benefits of "coming out." *Equity Excell. Educ.* **2004,** *37,* 4–17.

Vogel, S. A.; Leyser, Y.; Burgstahler, S.; Sligar, S. R.; and Zecker, S. C.; Faculty knowledge and practices regarding students with disabilities in three contrasting institutions of higher education. *J. Postsec. Educ. Disabil.* **2006,** *18,* 109–123.

Vogel, G.; and Sharoni, V.; My success as a teacher amazes me each and every day: perspectives of teachers with learning disabilities. *Int. J. Inclus. Educ.* **2010,** *15(5),* 479–495.

Wertheim, C.; Vogel, S. A.; and Brulle, A. R.; Students with learning disabilities in teacher education programs. *Ann. Dyslexia.* **1998,** *48,* 293–309.

Wolfensberger, W.; The Origin and Nature of Our Institutional Models. Washington, DC: Government's Printing Office; **1975**.

WrightsLaw. Summary of the Bartlett Case. Retrieved from http://www.wrightslaw.com/law/caselaw/case_Bartlett_Bar_2d_9809.html, **2013**.

Yanok, J.; Equal opportunity in teacher education programs for the learning disabled. *J. Teach. Educ.* **1987,** *38,* 48–52.

Zion, S.; and Blanchett, W.; (Re) conceptualizing inclusion: can critical race theory and interest convergence be utilized to achieve inclusion and equity for African American students? *Teach. Coll. Rec.* **2011,** *113(10),* 2186–2205.

CHAPTER 4

REDEFINING THE SOCIAL CONSTRUCTION OF COOL IN TERMS OF SOCIAL JUSTICE

LAUREN T. GONYEA

CONTENTS

Imagine a busy street in the early morning with cars passing by, filled with people rushing to get to work on time. Cars are able to travel freely, as there are no traffic lights or stop signs on the street. The only possible restriction of traffic is a white-lined crosswalk in the middle of the street. The sidewalks are empty, except for a young woman dressed in sweatpants and an old T-shirt, who is eagerly making her way to the gym. Cars zoom by as she approaches the white-lined crosswalk. As she assertively takes a single step into the road, the cars continue to speed past; drivers pretend that they do not see her. She is invisible to the drivers, and anxiously searches for eye contact with the one driver that will take the initiative to slow and allow her to safely cross the busy street. With no external traffic signals other than the white crosswalk, it is up to the drivers to acknowledge the woman in the crosswalk and allow her to safely cross. She takes another step forward and cars continue to speed by pretending that they do not see her. Twenty seconds pass and suddenly, a man dressed in a black pinstripe suit and dark sunglasses approaches the same crosswalk. The cars instantly slow and come to a stop before he even enters the crosswalk. Both pedestrians are safely able to make it to the other side.

I will never know why it took so much longer for the cars to stop for me that day. The man and I approached the crosswalk with the same amount of speed and assertion. Did the drivers truly not see me in the crosswalk, or did they feel that they could continue to drive and rely on the driver behind them to make the stop? Was it a matter of age, gender, or social class? As a young, middle-class, White, female, I cannot help but wonder how my identity influenced the decisions of the drivers. Did the man project a greater sense of power and importance than I had walking to the gym? Much like the drivers in this situation, mainstream society marginalizes minority groups.

The question now becomes thus: Would it have mattered if I was an older male, but still dressed in a sweat suit? What if I was a black male? What if I was a celebrity? If I was well known in the community, would that have changed the perspective of the drivers? Did I resemble any of the drivers' sons or daughters? If not, would they have been more likely to stop if I did? Beverly Tatum engages the topic of identifying with multiple identities. "When we think about our multiple identities, most of us will find that we are both dominant and targeted at the same time. But it is the targeted identities that hold our attention and the dominant identities that often go unexamined" (Tatum, 2010, p. 6). As educators, we must identify and analyze the privileges in our own lives. To do this, we must examine our dominant identities and actively push forward to prevent marginalization of minority groups. We must be the car that breaks the cycle, and slow to allow others to cross. Otherwise, underrepresented groups will continue to be thrown by the wayside.

4.1 DEFINING COOL

What does it mean to be cool? How does it relate to social justice? Dar-Nimrod defines cool in the article *Coolness: An Empirical Investigation.* "Insofar as coolness is a marker of apparently desirable characteristics, it also denotes observable criteria for social inclusion or exclusion" (Dar-Nimrod, 2012, p. 176). Consequently, people in minority groups are marginalized and socially excluded, and therefore are often not considered cool. Since cool is an indicator of social inclusion, it is an integral part of social justice work. However, educators will often step out of the conversation of peer culture, the social dynamic between peers (Corsaro, 1990, p. 200). In my own experience, my elementary, middle, and high school teachers have been reluctant to engage this topic with us. Milner (2006) notes that "Teachers and administrators tend to see peer culture as largely outside of their realm of responsibility—unless school rules or laws are clearly violated" (p. 186). That is why it is imperative that college educators and administrators help students to reset this default setting of society's perception of cool not only to help students reach their maximum potential but also to push toward social justice.

In the United States, cool is a reflection of perceived social class. In the book *The Nigger in You*, J.W. Wiley states, "Cool is understanding that the very notion of cool is a social class marker for accepted behavior that everyone on some level wants but no one wants to be seen pursuing" (Wiley, 2013, p. 90). By Wiley's definition, cool is not only about material possessions, but attitudes, actions, and behavior. It is related to knowledge and enlightenment. In her book, *We Real Cool,* bell hooks articulates the concept of cool in terms of social justice. "It is a fake stereotyped notion of cool, that denies the history of the "real cool," which was not about dissociation, hardheartedness, and violence, but rather about being intensely, connected, aware, and able to judge the right action to take in a given circumstance" (Hooks, 2003, p. 152). The history of cool originated from acute awareness of social justice issues. Without even realizing it, language, style, perspectives, reactions, and perceptions of those in American society are all products of cool. The words that leave the mouths of those who accept society's default perception of cool may not necessarily be the words of the individual, but the master narrative reflecting society's perception of cool. Therefore, it is essential for educators and administrators to analyze how perceptions of cool influence teaching and daily decision making.

In the story about the crosswalk, the steady flow of cars represents society's perception of cool. Only dominant groups that hold power are able to enter the crosswalk of privilege or cool. Dominant identities (able-bodied, White, heterosexual, upper-class, males) hold the power to determine what is cool. Targeted identities, or minority groups, are marginalized by default. Allan Johnson defines privilege in the article *The Social Construction of Difference*. "Privilege exists when one group has something of value that is denied to others simply because of the groups they belong to, rather than because of anything they have done or failed to do" (Johnson, 2010, p. 16). Thus, people with dominant identities are granted privileges that people with targeted identities are denied.

Individuals are perceived as cool not only by virtue of who they are in terms of their identity and their privileges but also by their actions as well. For instance, society may perceive someone possessing dominant identities who challenge the master narrative or social norms as uncool. In the book *Freaks, Geeks, and Cool Kids*, Milner (2006) recognizes the importance of deviance with systems that possess a high value on conformity. "Deviance must be persecuted lest it call into question the basic assumptions of the normative structure" (p. 90). Perhaps, this persecution is what prevents many of us from speaking out and becoming an ally. As educators, however, we must rise above this fear by challenging the social norms and attempt to change society's perception of cool.

4.2 CLASS AND COOL

One of the largest social justice issues in the United States is the American belief in the myth that the United States is a classless society. This is one of the myths that "our" education system engenders us to believe. One of my peers and I once had a debate on whether everyone has equal access to a college degree. My argument was that not everyone has the same privileges or opportunities that he and I had been granted throughout our lives. His argument was that if people worked hard enough, they would be able to achieve anything that they wanted to. He believed in the myth of meritocracy. This is an example of dominant groups reiterating the master narrative, or the "conventional wisdom" of the dominant group (Frost, 2012, p. 444). This often results in overused stereotypes.

When considering the struggles that people with lower economic class endure, the film *Frozen River* (Rae & Hourihan, 2008) comes to mind. In the film, the son offers to get a job to help his mother financially. She refuses his help and tells him he must stay in school. The father abandoned the family and the mother could not afford to pay the monthly mortgage payment with her part-time job. Since education is valued in our society, those without it often suffer financially. Garcia, Lardner, and Zeldin illustrate this struggle in the article *The Debt-for-Diploma System*, "But when students work longer hours at multiple jobs—and when they work off campus as well as on—the formula [students working to lower their debt] goes sour. Studying time dwindles (along with time for extracurricular and social activities), grades suffer, and the whole college experience takes on a harsher edge" (Garcia, Lardner, and Zeldin, 2010, p. 175). The combination of work and school quickly becomes overwhelming, and students with little or no financial support are left by the wayside. Upper-class students, who are not required to work multiple jobs to pay for their tuition, rent, or other various expenses, are able to afford to spend extra time and money on assignments and schoolwork. For instance, many of my friends could not afford to pay for prep classes to study for the scholastic assessment test (SAT) or revised general test (GRE). The cost of such courses begins around $300. The scores that students achieve on these tests determine not only their admission/denial into college, but how much financial aid they receive from the institution. Classes were available for those with extra time and money to help prepare for the exams. Educators subconsciously accepting stereotypes about people in the lower class, such as being lazy and neglectful, are doing a disservice to their students.

The film, *Boiler Room* (2000), offers an eye-opening perspective about the struggles and differences in social economic class. In the film, Jim Young (played by Ben Affleck) is a senior broker at J.T. Marlin. In a meeting with entry-level brokers, Young flaunts his wealth and status in order to appeal to the prospective brokers. Young promises that the applicants will make a million dollars within the first 3 years of working at J.T. Marlin. However, before the applicants are even able to make that much money, they must work endless hours, disregard any disapproval from friends and family, and survive on an entry-level wage ($150 a week). Young's behavior is supported by Taylor's perspective of class in *The Book of Cool*. "Simply telling people what they want is so much easier than predicting what they *might* want" (Taylor, 2009, p. 130). Young is well aware of the master narrative about economic class. Wealth is a marker of social inclusion. He knows that society respects the wealthy, and the concept of cool is often associated with wealth. Hence, Young flaunts the type of car he drives, his fluid assets, and his power at the firm. Young marginalizes values such as friends and family because he knows society does not hold those values to as high a standard as money or power. He warns the prospective brokers not to listen to friends and family because they will advise against the new lifestyle at the firm. Thus, he portrays the value of friends and family as uncool by dismissing their opinions, automatically rationalizing the dismissal with the fact that the prospective brokers will make more money than their families. He utilizes society's acceptance of dominant identities to his advantage.

In order for Young to maintain his social status, the pressure is on Young to find youthful brokers who are willing to work under such volatile conditions. Taylor urges in *The Book of Cool*, "The hunt is on, and the company that discovers the next cutting-edge trend will cash in big" (Taylor, 2009, p. 126). The main objective of the large firm is to train new and inexperienced brokers for very little money and bribe them with fame and fortune. Only a select few are chosen for the job and will enjoy the fruits of their labor. Those who are not chosen, however, are discarded, and the work that they completed up until that point goes to the benefit of the firm with little compensation.

> But the fundamental concern of corporations is to make a profit and this requires that they encourage employees to be good producers and the public to be eager consumers. Everything else is secondary. Their employees know this and those who wish to prosper organize their lives accordingly. (Milner, 2006, p. 169)

This mutual understanding allows big business to exploit employees. Educators should consider shining light on this type of exploitation in the workforce. The inexperienced brokers were blindsided by Young and were consumed by his portrayal of the concept cool. In school, we are taught about the abuse of factory workers during the Industrial Revolution. However, exploitation of workers still occurs today. Those at the top enjoy a comfortable lifestyle, whereas those at the bottom struggle to make ends meet. This is all masked by society's default concept of cool. The master narrative tells us that those who work harder will make more money and have more expensive possessions. This packaging of cool is something that educators should consider in order to protect students from entering potentially hostile work environments. Engaging students in the tough and often avoided conversations about the conventional values of society and current exploitation of marginalized groups is just one of many ways to initiate social justice work in the classroom.

4.3 RACE AND COOL

In my high school experience, more often than not, educators have tended to tiptoe around the issue of race. For most, it seems to be an uncomfortable conversation. People with dominant identities do not like acknowledging dominance and privilege. Author and educator Tim Wise urges in his essay *With Friends Like These, Who Needs Glenn Beck? Racism and White Privilege on the Liberal-Left* that in efforts to gloss over these conversations, there has been a movement toward colorblindness.

> What's more, when teachers and others resolve to ignore color, they not only make it harder to meet the needs of the persons of color with whom they personally interact, they actually help further racism and racial inequity by deepening denial that the problem exists, which in turn makes the problem harder to solve. To treat everyone the same—even assuming this were possible—is not progressive, especially when some are contending with barriers and obstacles not faced by others. (Wise, 2010)

By denying that racial inequality exists, marginalization is perpetuated, and we move farther from social justice. Consequently, educators must engage students in the conversations about social inequality.

In my first year of college, I was an elementary education major. In one of my first education classes, I had a professor who was in her first year of teaching at the college. She was generally very cheery and had previously worked as an elementary teacher. Her voice was particularly quiet for a teacher, and I often found myself wondering how rambunctious first graders responded to her demeanor. One day, we were having a discussion about celebrating holidays in the classroom. As an assignment, we were required to brainstorm activities that we could use for each holiday. The class suddenly became elated and came up with elaborate ideas for several of the popular holidays (Christmas, Easter, Columbus Day, etc.). Our class was generally an outspoken bunch so conversation flowed easily throughout the room. After about 5 min of collaboration, our professor raised a question to the class. She asked what we would do if we had a child in our classroom who believed in something other than Christianity. Some of my peers suggested that we continue to do the activity as planned. We proceeded to work on the assignment while she was speaking. Other students boldly shouted that the child go to the library and color, whereas the class continues the festivities. The class did not seem to understand why she would suggest "ruining the experience of an entire classroom" for only one person.

Usually, agreeable with our answers, a troubled look projected across our professor's face. Retrospectively, I can see that this situation posed a leadership moment for my professor. It was only her first year teaching at the college. At such a tight-knit college, the reputation of a professor is almost as important as the content being taught. She was not fully established in the system yet. A few of my peers had connections to administrators. In the previous semester, a few members of our class had sent letters to the dean asking for removal of two different professors. One letter resulted in the "removal" of a professor from our cohort. With such an outspoken class and only being her first year, she did not stand a chance if we turned against her. She could have easily backed away from this conversation in order to avoid confrontation with such an opinionated class.

In this moment, my professor calmly proceeded to probe the class by asking what we would do if some of the children in the classroom were Native Americans. A few of my peers engaged in whispered side conversations trying to figure out her angle. Interestingly enough, every person who engaged in side conversations was White. Thus, taking the onus off them and placing it on the person who was presenting the challenge. Predominately White, our class did not understand what our professor was trying to suggest. Was she trying to suggest that we were being insensitive to race? One student boldly suggested that they continue to do the activity as planned. After all, the students in our class seemed to agree that everyone must learn "the American culture." Suddenly, my professor broke free from any lingering cheerfulness in her body. With a burst of urgency and disappointment in her voice, her voice grew ominously. She demanded how we would explain celebrating Columbus in a traditional sense, to the child or the child's family. The room suddenly became eerily quiet and uncomfortable. All whispering and side conversations stopped. Our professor held onto this silence to emphasize its importance. For that moment in time, she possessed all of the power. I could not help but smile, finding an entirely new respect for her. We had never engaged in a conversation like this before, nor had we been challenged in such a direct fashion.

Our professor was challenging an idea that had been engrained into my subconscious since before I was born. We were in our first year of college. Why had not we been challenged on this idea of dominance before? Rodriguez suggests one of the contributing factors to the continued marginalization of invisible minorities is the lack of engagement in the educational standards. "In spite of the progress made in recent years, the current "standards craze" sweeping the country threatens the educational opportunities of already disadvantaged students by making invisible the social inequalities that gave rise to the calls for education reform and standards in the first place" (Rodriguez, 1999, p. 3). These conversations are often avoided because those who are not members of targeted groups are not comfortable engaging in discussions about dominance and privilege. Further, the topics not required by the standards will often remain untouched by secondary education teachers. Consequently, many educators consider this topic to be outside of the required content and can justify avoiding these topics. In many cases, dominant and privileged educators teach the way that they were taught. In other instances, dominant and privileged educators may neglect challenging the dominant culture in fear of persecution. Thus, avoidance of these types of conversations is perpetuated.

How does this story connect with the phenomenon of cool? More importantly, what can we take away from this in terms of social justice? Needless to say, my professor was not popular with the class after our discussion. Our class was predominantly White; consequently, the fact that we needed to learn how to let go of privilege was a hard pill for many to swallow. It was apparent that many were not even aware of their privilege. Several students became defensive, and judging by the look on my professor's face, she walked away that day beyond frustrated.

What my professor did not realize, however, is that in her leadership moment she was able to plant a seed. Her articulation not only catalyzed my interest in social justice issues but also my classmates. If educators want to articulately engage students in conversations about race, they must develop their use of language. For instance, an English professor simply cannot adequately engage students in discussions about Mark Twain's novel *Adventures of Huckleberry Finn* without developing a sophisticated lens about the issues of race and language. The class may not have fully understood where my professor was coming from at that moment in time, but the questions that she posed would linger in our future conversations. Throughout the year, the students slowly started to become more conscious of minority groups and the privilege that comes with being a part of the dominant group. That was only my professor's first year of teaching. She could have easily disengaged from her leadership moment in an attempt to establish herself as the new *cool* professor. Cool is dependent on perspective and context. At that time, the class perceived our professor as uncool because she challenged the belief of the dominant culture. Our class was predominantly White. By acting as an ally, our professor was challenging the belief of the dominant culture. Consequently, our class did not agree with this. At the time, our class possessed a traditional perception of cool. In many ways, I would argue that my professor was cool for providing the class with a different perspective and not backing down from her leadership moment. Wise also stated in his essay,

> [Its] embodiment is the elementary school teacher who I seem to meet in every town to which I travel who insists "they never even notice color" and make sure to treat everyone exactly the same, as if this were the height of moral behavior and the ultimate in progressive educational pedagogy. (Wise, 2010)

My professor in that class was trying to steer us away from the colorblind approach and attempting to engage us in conversations that were at the heart of the issue. By merely considering a different perspective, one can alter the perceptions of cool. Consequently, altering perspectives of cool is essential to combating marginalization and steering our society toward the direction of social justice.

4.4 ABILITY AND COOL

I recently worked as a Sports Director at a Cub Scout camp. Throughout the camp, I realized many of the kids looked up to me because I was a college athlete. I was an able-bodied female with a high level of energy and competitiveness. In their eyes, this was cool. These traits corresponded with social inclusion. During one of the sessions, I had a child in my group named "Eric" who was fond of trains. Throughout the sessions, he was adamant on carrying around his toy train with him during the activities. One day, we were playing kickball and Eric was holding onto his train in the outfield. For a second, I contemplated forcing Eric to put his train down in fear that he would fall on it while playing kickball. I resolved to let it go, and wait to see how he would adapt to playing kickball while holding his train.

It quickly became apparent that two of the other children were not content with Eric holding a train while playing kickball. "Andrew" and "Connor" believed that carrying around a train was uncool for a ten-year-old. Immediately, they started harassing Eric about his train. I overheard Andrew mocking Eric, calling him "retarded" while Connor attempted to snatch the train away from Eric. I quickly pulled Andrew and Connor aside and sat them down to talk. Judging by their expressions, I think they were expecting me to yell at them for their behavior. They were used to being yelled at by their leaders for their behavior.

I began the conversation by asking them what sort of names they dislike being called by others. They eagerly listed off a bunch of words. I asked them if they used those same words against others, and they agreed. Their eyes were wide; it was clear that they were curious as to where I was going with this. I told the boys that every time they use words such as "retard" they are hurting me. A puzzled look crossed both of their faces. Andrew immediately asked what I meant. He flat out told me that I am a college athlete; thereby, I am clearly not "retarded." Connor agreed. They did not understand how calling someone else "retarded" would personally influence me. I told the boys that I have family and friends who are differently-abled. Each time they use that word as an insult, they are insinuating that my family and friends who have disabilities are uncool. There was a moment of silence. I asked Connor if he would call Andrew a "retard" if he became severely injured playing kickball. Connor said no. I responded by indicating that we are all temporarily able-bodied and at any moment in time, we could become disabled. Using dysfunctional language not only hurts me, but them as well. I could instantly see that the boys were starting to understand the larger picture. Connor asked if using the words "fag," "homo," and "sissy" hurt me as well. I nodded. The boys resolved to apologize to Eric. To my surprise, they also apologized to me for their behavior. Throughout the rest of the week, I noticed that Andrew and Connor refrained from using the dysfunctional language that we previously talked about.

In a subtle way, I was able to influence Andrew and Connor's perception of cool. Slowly, the boys altered their behavior and language throughout the week. At the end of the week, I learned that Eric was autistic, and he used his train as a coping mechanism.

Disability is unique in the sense that it is one of the most inclusive minority groups. Approximately one in five Americans has a disability (US Consensus Bureau, 2012. Anyone can become differently abled at any time. Yet, why are students with disabilities among the first to be dismissed as uncool? Why would Eric instantly be targeted by Anthony and Connor? Is this because able-bodied people can see the potential in themselves to be differently abled? Is that why we want to exclude those who are differently abled and create a larger barrier between ourselves? Milner (2006) provides insight into potential motivation of the marginalization of minorities.

> First, in an atmosphere where put-downs and verbal aggression are common, there is probably a definite tendency to displace hostility by scapegoating the vulnerable. This mechanism has been widely alluded to as a reason for the persecution of minority groups. Second, in a context where verbal aggression is common and even admired, the vulnerable offer an opportunity to hone and display one's skills without risking significant retaliation. Third, in a system where the content of many of the norms is obviously highly arbitrary—as in the case with fashion and style—an even higher value is placed on conformity. (Milner, 2006, p. 90)

Since many students with disabilities deviate from the social norms, they are persecuted by society. Thus, Milner is suggesting that there are three contributing factors that promote the marginalization of minorities in classrooms.

The first contributing factor that promotes persecution of minorities in the classroom is done by promoting a classroom environment in which verbal put-downs are common. The second is promoting a classroom environment which encourages verbal hostility and providing few consequences for these comments. Either of these can be done by choosing to ignore hostile comments or seemingly innocent jokes. In the story of summer camp, had I ignored Andrew and Connor's taunting comments, the harassment of Eric would have been further perpetuated. Alternatively, perhaps I subconsciously ignored comments that should have been addressed before I intervened on Connor and Andrew blatantly harassing Eric. Finally, Milner suggests that placing a high value on conformity also contributes to the marginalization of minorities. For instance, Eric carrying around his train was an example of a behavior that did not conform to societal norms. Hence, he was persecuted for his actions. Had I forced Eric to put down his train while playing kickball, I would have further encouraged conformity and created an even less inclusive environment. Eliminating dysfunctional language, changing the value of conformity, and creating inclusive environments are some of the many ways of helping students who are differently abled.

Dar-Nimrod illustrates the link between society and cool. "The popular essence of what is considered cool today may simply be what most people think they are supposed to find desirable" (Dar-Nimrod, 2012, p. 177). Thus, altering what people are supposed to find desirable is essential to altering the perception of cool. Having a conversation with Andrew and Connor about the use of dysfunctional language was much more productive than simply telling them that they should not use those words. They were accustomed to being yelled at. Educators have the power to change the perception of cool and guide students towards social justice. Educators have the power to influence students' perception of cool and stand up for social justice. I believe Eli Clare's articulation in the article *Gawking, Gaping, Staring* says it best.

> Cripples, queers, gimps, freaks: we are looking for lovers and teachers—teachers to stand with us against the gawking; lovers to reach beneath our clothing, beneath the words that attempt to name us, beneath our shame and armor, their eyes and hands helping return us to grace, beauty, passion. (Clare, 2010, p. 496)

4.5 SEXUAL ORIENTATION AND COOL

The lesbian, gay, bisexual, transgender, and questioning (LGBTQ) community struggles to gain social justice as an invisible minority. Many heterosexuals feel the same way as David Burns, an administrator at Rutgers University, once felt.

> I don't tell everyone about my sexuality, so why do some gay men and lesbians make such a big deal about theirs? I wondered—and I was not alone in wondering—why we had to have all these gay dances, gay pride week, "wear blue jeans if you are gay" days. I remarked to myself about the "stridency" of this form of politics. I guess I would have preferred silence. (Burns, 2000 p. 307)

Many faculty and administrators prefer silence because it is less controversial. However, silence eradicates the voice of the marginalized. Beverely Tatum relates racism to the moving walkway at an airport. "Passive racist behavior is equivalent to standing still on the walkway. No overt effort is being made, but the conveyor belt moves the bystanders along to the same destination as those who are actively walking" (Tatum, 2010, p. 68). Her description of racism can be analogous to any aspect of bystander behavior in terms of social justice. Anyone who remains passive on the moving walkway engages in unearned privilege that can result in homophobia. Often, dysfunctional language is able to penetrate protective shields, but does not show on the faces of the victims. Eventually, hurtful words may be internalized or redirected. The teammates on my college soccer team are frequently called "fag," "butch," and "homo." My friends who are homosexual are socially excluded or marginalized by society through this use of language. The default perception that homosexuals are uncool is reiterated by derogatory use of these hurtful words. Over time, hard feelings build up from the dysfunctional language. Consequently, I have witnessed many of my friends get into fist fights with others who use this type of dysfunctional language. This is just one story of what occurs outside of the classroom. As students enter the classroom, they bring this background with them. However, most educators are not aware, let alone acknowledge this. One of the privileges of being heterosexual is not being forced to engage in conversations of how the members of the LGBTQ community are marginalized.

Often times, those in the LGBTQ community are forced to hide or suppress their true identity. In the animated series *Boondocks*, the character Gangstalicious is a famous rapper who believes that his friend Riley witnessed him kissing another male. Consequently, he asks Riley not to tell anyone that he is gay. Riley does not understand this request, as he does not remember seeing Gangstalicious kissing another male. However, Riley idolizes Gangstalicious and when others question Gangstalicious' sexuality; Riley entirely denies the possibility of Gangstalicious being gay. At first, Gangstalicious thinks Riley is being cool and understanding. However, at the end of the episode, Riley confronts Gangstalicious and asks him if he is gay. At this point, Gangstalicious denies being gay. In fact, Gangstalicious says that Riley calling him a homosexual is gay. In *The Book of Cool*, Taylor illustrates Gangstalicious' motive. "Homophobia, of course, still exists, but it is no longer cool to call someone else a fag, unless, of course, you are gay yourself" (Taylor, 2009, p. 47). Gangstalicious takes it a step further by saying how disgusting it is to kiss another male and how he needed to brush his teeth after being kissed. In this instance, Gangstalicious is dabbling in W.E.B. DuBois' idea of double consciousness. "It is a peculiar sensation, this double-consciousness, this sense of always looking at one's self through the eyes of others, of measuring one's soul by the tape of a world that looks on in amused contempt and pity" (DuBois, 2008, p. 12) Gangstalicious is conflicted between society's perception of being a stereotypical black male and a homosexual male.

Many times, people who are marginalized use dysfunctional language toward members of their own group. This is because according to society, it is socially acceptable to use dysfunctional language against your own identity group. Thus, society promotes internal marginalization in order to further suppress minority groups. What does this mean for higher education faculty and administrators? Educators must understand that many of their students endure internal marginalization and engage in double consciousness on a daily basis. On the surface students may appear to be fine, but it is only a protective shield veiled by the phenomenon of cool. Considering Tatum's (2010) perspective of the moving walkway, educators have four choices. The first choice is to actively walk with the moving walkway, thus acting as the persecutor. The second choice is to remain passive on the moving walkway, still further perpetuating the marginalization of minority groups. Those who remain passive on the walkway are heading for the same destination as the active walkers. The third choice is to face the opposite direction on the moving walkway and turn your back against homophobia. However, unless educators actively begin walking in the opposite direction at a pace that is faster than the conveyor belt, they will find themselves at the same destination as the others. This can be done by engaging students in conversations about the power of privilege, language, and dominance. The final choice is the only selection in favor of social justice.

4.6 GENDER AND COOL

I started playing soccer when I was 5 years old. The support of my parents has granted me the privilege of playing ever since. Today, I am a collegiate soccer player entering my final season of eligibility. Each summer, I play on coed soccer teams. In these games, each team plays with eight people on a field. Since it is coed, there is a rule that each team must have a minimum of two females on the field at all times. Many teams struggle to get enough players due to various schedule conflicts. The only issue that I have found on these coed teams is that females are often treated as "cones" on the soccer field by the male players. That is why many believe that substituting a male for a female is equivalent to substituting a male for a motionless cone, consequently putting your team at risk of losing the game. Ironically, "this strategy" backfires as the best teams in the league are the teams that utilize the female players to the fullest. Mathematically speaking, this makes sense. The best teams in the league pass to eight players on the field, rather than six. Unfortunately, many of the teams in the league have yet to pick up on this.

Recently, I volunteered to play on a coed team that was in need of players. They had never seen me play before and I had no prior connections with any of the players. We did not have any substitutes and were barely able to scrounge up enough players to play. Before the game even began, one of the males cautioned that if another female player showed up, she needed to rotate with one of the other females on the field. This is a common suggestion. Many of the males in the coed league have a hard time substituting with females. They prefer to utilize a maximum of two females rather than a minimum of two. As the game began, one of my teammates instantly began to bark instructions to me on the field. Communication on the soccer field is essential; however, I could sense that this male's tone was condescending. He was speaking to me as if I had never played the game before. It would have been different if he had communicated with the rest of the team like this. Yet, he was only giving these basic orders to the other female and me. I found this to be irritating. However, I never said anything. I was used to this. Often when playing on coed teams, males refuse to pass to females. Since females are treated like cones, males prefer to try and dribble through four people than pass to a female who is wide open. This is exactly what happened in this game. It got to a point where I wanted to walk off the field because I was only passed to once, even though I was wide open in almost every play. I was not even given a chance.

Before the halftime whistle blew, we were down by a single goal. At halftime, I exchanged positions with one of the other males. He was starting to get tired so he asked if I could switch with him and play center midfield. Five minutes into the first half, I battled for the ball against one of the males on the opposing team. I won the battle and immediately headed to goal. Taking a left-footed shot, the ball soared into the top corner of the goal. The referee told me that he had not seen a goal like that in a long time. After that play, the males on my team were in awe. The male who had been talking down to me instantly stopped. He suddenly changed his tune and started telling the male players to not only pass to me, but the other female as well! This correlates to Wiley's definition of cool. "Cool is when we know we have made a mistake, owned it, and went about correcting it" (Wiley, 2013, p. 5). Once I scored, my teammate realized his bias and started making amends. The moment he told the other males to start passing to me, I was able to contribute three additional assists and we were able to walk away with a win.

In the heat of the moment, people resort to their embedded biases. The way I was marginalized in the game is reflective of the way in which females are marginalized in society. Unequal pay in the workforce, lack of representation in Congress, and exploitation of the female body in the media are all examples of how women are marginalized in society. On the soccer field, in the work force, in the classroom, and in any aspect of life that has pressures to succeed, people are presented with opportunities to show their true colors. We must identify and analyze our own biases to figure out who we are subconsciously marginalizing. It is human nature to use stereotypes. However, we must step out of our comfort zone and change our default responses. Otherwise, we are merely products of our society. There were six males on the field. All it took was one to step up to literally change the game. However, I should not have had to prove myself to gain the respect of the other players. The true measure of cool would have been for one of the males to speak up before I scored the goal. This story illustrates that acknowledging, accepting, and embracing differences is not only essential to steering our society toward social justice, but enriching the fabric of life.

The phenomenon of cool is a powerful tool that society uses to marginalize others. Relative to social justice, it is used in terms of class, race, ability, sexuality, and gender. Educators must understand their influence in higher education and seize the opportunity to change the default perceptions and stereotypes of cool. In my own life, I know stepping up as an ally by overcoming my own biases and prejudices is essential. However, I believe the true test of cool is the ability to give up privilege in efforts to support a total stranger. To adjust the social imbalance, people with dominant identities must acknowledge and give up many of the privileges of being in the dominant group. One of the privileges of dominance is unawareness of dominance itself. Amongst many things, educators should acknowledge their own biases and privileges. Once these biases and privileges are acknowledged, educators can begin to provide a lifeline to students who are often subconsciously dismissed or neglected. Much like the story of the crosswalk, educators must actively consider their students who are anxiously waiting to cross the street.

KEYWORDS

- **Cool**
- **Gender**
- **Race**
- **Social justice**

REFERENCE

Burns, W. D.; Why do not gay people just keep quiet? Listening to the voices of the oppressed. In: Eds. Adams, M.; Blumenfeld, W. J.; Castaneda, R.; Hackman, H. W.; Peters, M. L.; and Zuniga, X.; Readings for Diversity and Social Justice. New York: Routledge; **2000,** 306–309 p.

Clare, E.; Gawking, gaping, staring. In: Eds. Adams, M.; Blumenfeld, W. J.; Castaneda, R.; Hackman, H. W.; Peters, M. L.; and Zuniga, X.; Readings for Diversity and Social Justice. New York: Routledge; **2010,** 495–498 p.

Corsaro, W. A.; and Eder, D.; Children's peer cultures. *Ann. Rev. Sociol.* **1990,** *16(1),* 197–220.

Dar-Nimrod, I.; Hansen, I.; Proulx, T.; Lehman, D.; Chapman, B.; and Duberstein, P.; Coolness: an empirical investigation. *J. Ind. Diff.* **2012,** *33(3),* 175–185. doi: 10.1027/1614-0001/a000088

DuBois, W. E. B.; The Souls of Black Folk. Rockville, MD: Arc Manor; **2008.**

Frost, J.; Using "master narratives" to teach history: the case of the civil rights movement. *Hist. Teach.* **2012,** *45(3),* 437–446.

Garcia, J.; Lardner, J.; and Zeldin, C.; The debt-for-diploma system. In: Readings for Diversity and Social Justice. Eds. Adams, M.; Blumenfeld, W. J.; Castaneda, R.; Hackman, H. W.; Peters, M. L.; and Zuniga, X.; New York: Routledge; **2010,** 174–178.

Hooks, B.; We Real Cool: Black Men and Masculinity. New York: Routledge; **2003**.

Johnson, A. G.; The social construction of difference. In: Readings for Diversity and Social Justice. Eds. Adams, M.; Blumenfeld, W. J.; Castaneda, R.; Hackman, H. W.; Peters, M. L.; and Zuniga, X.; New York: Routledge; **2010,** 5–20 p.

Milner, M. M. Jr.; Freaks, Geeks, and Cool Kids: American Teenagers, Schools, and the Culture of Consumption. New York: Routledge; **2006**.

Rae, H.; Hourihan, C.; (Producers), and Hunt, C.; (Director). Frozen River [Motion Picture]. United States: Sony Pictures Classics; **2008**.

Rodriguez, A. J.; Making ethnicity invisible in the name of equity: standard contradictions in the national science education standards. *Multicultural Perspectives,* 1999, *1(2),* 3–7. doi: 10.1080/15210969909539893

Tatum, B. D.; Defining racism: "Can we talk?" In: Readings for Diversity and Social Justice. Eds. Adams, M.; Blumenfeld, W. J.; Castaneda, R.; Hackman, H. W.; Peters, M. L.; and Zuniga, X.; New York: Routledge; **2010,** 5–8.

Tatum, B. D.; The complexity of identity. In: Readings for Diversity and Social Justice. Eds. Adams, M.; Blumenfeld, W. J.; Castaneda, R.; Hackman, H. W.; Peters, M. L.; and Zuniga, X.; New York: Routledge; **2010,** 5–8.

Taylor, M.; The Book of Cool. Philadelphia: Running Press; **2009**.

Todd, J.; Todd, S.; (Producers), and Younger, B.; (Director). Boiler Room [Motion Picture]. United States: New Line Cinema; **2000**.

US Consensus Bureau. Nearly 1 in 5 People have a Disability in the US, Consensus Bureau Reports. Retrieved November 10, 2013, from http://www.census.gov/newsroom/releases/archives/miscellaneous/cb12-134.html, **2012**.

Wiley, J. W.; The Nigger in You: Challenging Dysfunctional Language, Engaging Leadership Moments. Sterling, VA: Stylus Publishing; **2013**.

Wise, T.; With Friends Like These, Who Needs Glenn Beck? Racism and White Privilege on the Liberal-Left. Message Posted to http://www.timwise.org, August 17, **2010**.

CHAPTER 5

HOW SELF-KNOWLEDGE IMPACTS THE ABILITY TO TEACH ETHICALLY: TEACHING JOURNEYS

ALINE R. BOBYS

CONTENTS

5.1 TEACHING AS A MORAL PRACTICE

> *The future isn't something hidden in a corner. The future is something we build in the present—Paulo Freire*

What would you be willing to give up in order to make someone else's life better? How are self-knowledge and an understanding of personal responsibility foundational for making teaching and learning decisions that are in the best interest of all? How do you advocate for others through collective action? And, finally, what is the role of literacy in democratic and global societies?

Graduate students are challenged to answer these questions and many others as part of their work in becoming literacy specialists in the M.S. Ed. Literacy Education program at SUNY Plattsburgh. A strand that runs through this program is multicultural competencies, with an emphasis on issues of social justice. Our philosophy of literacy instruction recognizes that literacy is a complex, interactive, constructive process and that differences among learners, with respect to cultural, linguistic, ethnic, and socio-economic diversity, exist in all classrooms.

We begin our program at the foundational level in EDR513: *Literacy for Democratic and Global Societies.* Students in this class are faced with rethinking the traditional definition of literacy as "reading and writing" and are asked to consider how twenty-first century definitions of literacy are tied to social, political, and cultural processes of globalization. They investigate how language, literacy, ethnicity, race, gender, and other social and cultural constructs affect how we communicate in a diverse democratic society, engage in social practices, live in relationship with each other, and contribute to justice and democracy. They explore how a literacy education that fosters communication, interaction, and collaboration with others leads to a thoughtful, committed, and active citizenry.

Dewey (1933) believed in creating environments for students to think independently and integrate their ideas into their learning. In doing so, teachers create living communities that give students time to think and discover. This lived community experience provides students an entry point to consider the needs of others and ultimately become participatory citizens in our democracy. How do teachers achieve this in a classroom when they themselves are not in environments that are nurturing and thoughtful? When they have not experienced a truly democratic classroom themselves? Our journey in this class is what grows possibilities.

Graduate students in EDR513 are challenged to live and work in a democratic classroom where their collective effort promotes an understanding of democratic education and their roles in becoming education reformers and advocates for students. They explore the ethical dimensions of teaching as well as their personal teaching values. They engage in collaborative and individual community action projects that emphasize literacy, community, democracy, and social justice. They live the messiness of teaching, the uncertainty of creating democratic classrooms, the challenges of decision-making and problem-solving, the barriers to understanding, and yet do so within a celebration of the personal worth of each other. This chapter focuses on our journeys.

5.2 THE BEGINNING

People, like all life, only change when they allow an event or information to disturb them into voluntarily letting go of their present beliefs and developing a new interpretation. Nothing living changes until it interprets things differently. Change occurs when we let go of our certainty-our beliefs and assumptions-and willingly create a new understanding of what's going on—Margaret Wheatley

Our literacy education faculty redesigned the M.S. Ed. Literacy Education program. Inspired by the works of Dewey (1916; 1927/1954; 1933), Freire (1970), Banks (2003; 2004), Ayers (2010), Beane (2005), and others, we made many changes to our courses and our direction was clear: we wanted to help our students become literacy leaders who had agency and, with that agency, a commitment to advocacy and social justice for all students. In their roles as literacy specialists and literacy coaches, they would have the skills to guide those with whom they worked to effect change within all aspects of their school and community. Banks (2003) notes that "A literacy education that focuses on social justice educates both the heads and hearts of students and helps them to become thoughtful, committed, and active citizens in their nation and the world" (pp. 18–19). To achieve this goal of participatory democracy through education, we wanted to help our students understand Freire's concept of "liberating education"—humanization—where all in a classroom are teachers and learners. Freire (1970) argued that "education must begin with the solution of the teacher-student contradiction, by reconciling the poles of the contradiction so that both are simultaneously teachers and students" (p. 72). How do you learn this if you do not live it?

As the architect of one of the newly created foundational courses, *Literacy for Democratic and Global Societies,* I was selected to teach this course for its first offering and needed to prepare my course syllabus. I started to frame the usual information: course title, description, prerequisites, etc. What next? Typically, I would create assignments, assign points, and create a schedule of readings and due dates, but I was unusually uncomfortable doing so for this course. I kept reading and rereading my own words in the course description. How could I invite my students to live in a democratic classroom if I was making all of the class decisions before I had even met them? I was stuck. I realized I needed to change, and change meant letting go of my previous notions of controlling the planning of a graduate class for my students. The decisions I needed to make for this class needed to be based on what I believe "about the purpose of schools, the role and rights of students, and [my] obligation as a professional educator" (Beane, 2005, p. 2). If I believed that the way to learn participatory democracy and its connection to social justice teaching was to actually live the democratic way, my planning had to reflect that. "At its core, democracy rests on three fundamental principles—respect for human dignity, concern for common good, and faith that human beings working collaboratively are capable of resolving issues that confront them" (Beane, 2005, p. 119). Learning about democracy, social justice, literacy, and community in relation to teaching as a moral practice could not come out of merely reading texts about them. Therefore, I decided to start with what I wanted my students to know and be able to do by the end of the course. That led to the creation of enduring understandings and essential questions that follow.

5.3 ENDURING UNDERSTANDINGS

Understanding 1: Self-knowledge and personal responsibility are the foundation for making teaching and learning decisions that are in the best interest of all.

Understanding 2: Language, literacy, ethnicity, race, gender, and other social and cultural constructs affect how we communicate in a diverse democratic society, engage in social practices, live in relationship with each other, and contribute to justice and democracy.

Understanding 3: Teachers have an ethical responsibility to meet the needs of all students.

Understanding 4: Theoretical constructs and pedagogical practices are interrelated.

5.4 ESSENTIAL QUESTIONS

- What is the purpose of education? What is worth learning? How should debates over these issues be resolved in a democratic society?
- What makes education in a democracy unique? What counts as evidence of that uniqueness?
- What roles do schools play in educating citizens in a democracy for a democracy?
- How do our practices in schools foster democratic life?
- What would "education as the practice of freedom" look like? What policies would foster such an educational system?
- How do different school-based reform strategies attempt to improve education? What assumptions do these strategies make about the causes of educational inequality? How are reform strategies related to the purpose of education?
- How do different theories attempt to explain racial, social class, and gender gaps in educational achievement?
- How do educators construct curricula that link rich outcomes, literacy strategies, community activities, evaluation, and social change? What are the consequences when a student's needs are not met?
- How do cultural beliefs, social organization, and political contradictions shape teacher–student interactions in the classroom?
- In what ways are teachers "theory-makers"? How do teachers use theory and their knowledge of pedagogy to inform instruction?

Now what?

Maxine Greene (1995) states thus:

> Our classrooms ought to be nurturing and thoughtful and just, all at once; they ought to pulsate with multiple conceptions of what it is to be human and alive. They ought to resound with the voices of articulate young people in dialogues always incomplete because there is always

> more to be discovered and more to be said. We must want our students to achieve friendship as each one stirs to wide-awakeness, to imaginative action, and to renewed consciousness of possibility. (p. 43)

That is what I wanted my class to look like. I was determined to create a democratic classroom where our teaching and learning community would go from *me* to *we*. That meant I needed to remember my students were already classroom teachers and their experiences were going to inform our understandings. I needed to remember my students were adults and were very capable of making decisions, and I wanted them to have the opportunity to do so. I knew taking ownership of this class would be scary for my students because, as Daniels and Bizar (as cited in Beane, 2005) note, "The 13 years of submission and passivity customarily provided to young people by American schools is an exceedingly poor preparation for resourceful, self-initiating problem solvers, not to mention free and critical citizens" (p. 81). I also needed to remember that as a community of learners, my class would lead me to what they needed; navigating the ambiguity I knew they would experience in making decisions about their own learning would require some guidance.

5.5 THE INVITATION

> *Every effort must be made in childhood to teach the young to use their own minds. For one thing is sure: If they don't make up their own minds, someone will do it for them—Eleanor Roosevelt*

I finished the syllabus, but there was not much on it beyond the required texts, enduring understandings, and essential questions. What I left for my students was a scant frame from which they would build their class. I wanted them to have the experience of living in a classroom community where they had the right to and the opportunity for self-governance. I believed that the best of their teaching values would surface, and that they would be able to generalize this experience to creating democratic classrooms and communities in their own schools. I wanted them to see that in being given an opportunity to take responsibility over their own learning, they would have high expectations of themselves and each other and their work would have meaning. I wanted them to see that learning happens outside the four walls of a classroom and that experiencing democracy means having the life skills to make the world a better place through the ethics of caring. Learning how to work collaboratively would mean maintaining a community where diverse opinions are valued, where all voices are heard, where we ask a lot of questions that lead to even more questions, and where we help each other access innumerable resources and ideas. In every situation in our classroom, I wanted us to collectively and individually "consider the ethics of situations, issues, and events" (Beane, 85); I wanted my students to discover that these life skills would generalize to their roles as classroom teachers or literacy coaches, and that they would be able create this same democratic learning environment with their students and colleagues.

Deborah Meier (as cited in Beane, 2005) notes, "The solution to the messiness of democracy is more of it—and more time set aside to make it work" (p. 105). My classroom was going to be messy, and my students were going to be uncomfortable, but I needed to set aside the time to make this work. I invited my students into this process on the first night of class; I gave them three parameters around the invitation, and told them they were going to make all decisions concerning their semester learning around those parameters. The first parameter required them to have a reading partner in class to whom they would write reflections over their readings. I provided the book list, but the class was going to determine what to read, when to read it, who their partner would be, and when their reflections to each other were due. Their reflections were going to be the catalyst for classroom discussions, which they would lead. For the second parameter they were, as a class, to collaborate on a social justice project that contained the four elements of literacy, community, democracy, and social justice. They had to take this collective action project out of our classroom and they had to make it public, but they would be given class time to work on it. The final parameter was an individual social justice project that held special significance to them and that would also include the core elements of literacy, community, democracy, and social justice. They would have to make decisions both individually and collectively. Those who were naturally leaders would have to invite others to lead. Those who typically held back would have to step up and take a risk. All voices had to be heard.

Panic

"But how will we be graded?" Fourth parameter—as part of your experience this semester, you will create the grading criteria and rubrics for your projects; the partnership project criteria will be a class decision, the collaboration project criteria will be a class decision, and the individual project criteria will be an individual decision. They accepted the invitation.

5.6 BUILDING COMMUNITY

> *As we create communities from the cohering center of shared significance, from a mutual belief in why we belong together, we will discover what is already visible everywhere around us in living systems. People's great creativity and diversity, our desire for contribution and relationships, blossom when the heart of our community is clear and beckoning, and when we refrain from cluttering our paths with proscriptions and demands. The future of community is best taught to us by life— Margaret Wheatley*

It is not enough for students to read about the impact of learning in a democracy; they have to experience it. If I want my students to create democratic classrooms, they have to live in one. Building a democratic classroom community with students who have no experiences from which to draw is challenging and often requires deconstructing past learning environments. It also often meets with resistance.

Collaboration and cooperation play a central role in building community and in creating a space for students to *own* their learning as well as for respecting the learning of others; out of building a strong classroom community comes trust, shared values and responsibilities, and common goals. Building relationships is a main focus for every class meeting and is at the heart of the process. In my experience, when people are in relationship with each other, they hear each other, support each other, and trust each other, even if they disagree with each other. My schooling in Johnson and Johnson's (1989) model of cooperative learning and research by Furman and Shields (2003) on the concept of a creating a democratic community informed my thinking around creating this environment. According to Furman and Shields (2003), a democratic community includes the following:

- Respect for the worth and dignity of individuals and their cultural traditions

- Reverence for, and proactive facilitation of, free and open inquiry and critique
- Recognition of interdependence in working with the common good
- The responsibility of individuals to participate in free and open inquiry
- The importance of collective choices and actions in the interest of the common good. (p. 9)

This concept blends nicely with the goals of cooperative learning, which include an emphasis on positive interdependence, critical thinking, group processing, and social skills development. Building stamina in these areas is the bridge from the classroom to the global community.

Community building in our class is based on several consistent elements. First, we engage in some type of community builder at the beginning of each class. I facilitate the community builder the first week and each of my students takes responsibility for leading a community builder for each of the weeks after that. Community builders generate positive energy and help create a bond among us so that we can communicate authentically, deepen our relationships, and build trust. Second, our class sits *with* each other, in circular fashion, where the students and the teacher are equal participants in class meetings and conversations. In sitting knee-to-knee and eye-to-eye, we build positive interdependence and practice listening skills, pragmatics, and social skills. No one is allowed to sit outside of the circle. The first night of class we discuss how to have crucial conversations, where everyone's voice is heard and then we process, or debrief, at the end of each class to determine how these conversations went and how we might improve them for the next class. Next and most importantly, students work together to make decisions about building their semester course experiences around the parameters set forth. For example, students are given a text set from which they determine what they will read, how much they will read, and by what date they will make an initial response over the readings to their reading partner. Because this experience requires independent reading, writing, and reflecting, as well as having an online conversation with their reading partner, students must also determine by when they will give a response to their reading partner in order to be ready for class discussion the following week. Reading partnerships, although they happen outside of the classroom setting, foster community and explore democratic practices in addition to engaging students in a critical comprehension strategy. The result is that students form connections with their classmates while, at the same time, developing a deeper understanding of the readings and a correlation of these ideas to their own lives. In addition, because students are charged to lead class discussions with their reading partners, they feel responsibility toward each other and to the group. Johnson and Johnson (1989) often liken this concept of positive interdependence to *sinking* or *swimming* together. The responsibility students feel toward each other emerges in almost all of their final reflections, and students point to this experience as one of the most meaningful of their semester. For example, Meg reflects,

My literacy partner and I benefited from one another, for we constantly sprouted new ideas and built on old ones to promote the four themes of our course. One of the most important experiences I gained from working with a partner was the knowledge that she brought to the readings. Because of her, I still yearn for more knowledge about concepts we discussed in order to benefit both my students and myself. (EDR513 Final Reflection, Fall, 2011)

In another example, Karyssa notes,

The activity of participating in an online discussion partnership demonstrated the use of democracy as well. Sarah and I were both free to choose what we wished to discuss from our readings and were able to decide how we wished to present the information that we learned. Reading each other's posts allowed us to see a different person's point of view and discuss the issues that the other presented. (EDR513 Final Reflection, Fall, 2011)

Why is community building so important in a democratic classroom? What I have learned about community building from my students is that it promotes a culture of inquiry "framed by questions and fed by wide-ranging information" which "eventually leads to questions of equity and justice. When people raise questions about significant issues, they almost always end up asking why things are the way they are" (Beane, 2005, p. 76).

Students are also given the task of creating, developing, and implementing a group collaboration project as well as an individual project, and the criteria for those projects are developed within the group in myriad conversations with each other. This decision-making process is a dynamic one; often students will make a collective decision and then circumstances or additional thinking by fellow group members may mean a change needs to occur. My role is to help guide this process and model flexibility, always with an eye on how my students might generalize this process to their own classrooms.

Empowering students to make decisions around their learning promotes critical thinking. Each semester, decisions about how *Literacy for Democratic and Global Societies* will be run are made by the students, decisions around the group collaboration project are determined by the students, and how they will make public and assess the project is determined by the students. If I believe children should be able to explore their wonderings and become critical thinkers and problem solvers, then guiding their teachers in these practices is crucial. Engel and Martin (2005) note, "… children—all human beings in fact—are capable of making responsible choices, being engaged by things of the world, asking good questions, and becoming independent thinkers and learners" (p. 17).

Although I believe giving my students ownership over their learning gives them a sense of freedom, at the beginning of the semester, this freedom makes them uncomfortable and thrusts them into a state of disequilibrium. My students are from the "just tell me what to do and I will do it" generation of students, therefore living in a classroom where they have to make the decisions that impact their learning, and which to them translates to impacting their grade, is a new and scary experience. As Sarah noted,

> In the beginning of class I felt that having voice and choice within the classroom and sustaining a democratic classroom was near impossible. The idea that students could be involved in the decision-making was intriguing and frightening at the same time.
>
> (EDR513, Final reflection, Fall, 2011)

Teri felt similarly when she wrote,

> Many educators are skeptical of adopting this democratic way of life as they fear the unknown and the shift of authority. Initially, I found myself intrigued yet threatened by the components of this style of teaching...I was rather intimidated. (EDR513, Final Reflection, Fall, 2011)

Finally, Amber summed up her feelings when she wrote,

> At first the idea of a democratic classroom was just plain scary. Scary because of its obvious differences from the current teaching methods. I felt unsure about the ideas in this approach and how to incorporate them in my teaching. I was skeptical of their place in education today. (EDR513, Final Reflection, Spring, 2010)

My students have misgivings; despite their misgivings and skepticism, I ask my students to trust me and to trust the process. But why should they? They tell me that they have had plenty of teachers who talk about the concept of democracy in classrooms, progressive education, and social justice, but then do not themselves engage in those practices. What I have learned from my experiences in teaching is that, for students, trust has to be earned. You cannot just say "trust me" and have it happen. You have to mean what you say; you have to keep your word. You have to have high expectations of your students and, at the same time, let them know that you trust who they are as learners. In other words, you have to create the environment and within that environment, multiple opportunities for them to succeed. You have to support and cajole, and then get out of their way.

5.7 THE MIDDLE—GROUP COLLABORATION AND INDIVIDUAL PROJECTS

> *Paulo Freire once reminded us that, "whoever teaches learns in the act of teaching, and whoever learns teaches in the act of learning" (as cited in Cherland and Harper, 2007, p. 33). I think that this quote so perfectly describes the circle of teaching and learning that took place in our classroom every Tuesday in Sibley Hall. A classroom that for us, was not just a room, but a place where we were able to unite in our thinking, a place where we were able to teach each other through our learning and learn from each other through our teaching, a place where we were able to bloom and grow—Logan Gravel, Final Reflection*

As the semester progresses, students begin to embrace the idea of taking on more and more of the decision-making as they plan their collaborative and individual social justice projects. Even with all of the community building and trust building activities we engage in, there are still remnants of skepticism as to whether a group project is really possible and whether what they agree on will make a difference. By this time in the semester, students are convinced they want to do something together that is meaningful and purposeful to them as individuals and as teachers. I ask them to consider what they would do if they could make a change that would improve someone else's life. This challenge has presented itself in every semester in which I have taught this class. The most challenging aspect of group collaboration, of course, is where to begin, especially with so many voices in the room that need to be heard. Catherine reflects,

> At the beginning of the semester I wondered how 8 strangers would be able to collaborate on such a large project. How could all of us agree on one specific topic and meet all the requirements of the class? Little did we know how close together it would bring all of us. (EDR513, Final Reflection, Spring, 2011)

Michael spoke to the challenge of this process when he said,

> This project began with rough discussions of what we should do. We had difficulty narrowing our ideas for some time, but I believe this was because this group as a whole was not used to working as a collaborative group. Only when each and every voice in the group is allowed an equal say in the process will you truly have a complete group project where everyone has ownership in the outcome. This does not come naturally; I feel it is something that has to be worked on and practiced to be effective. (EDR513, Collaboration Paper, Spring, 2010)

He was correct. Where in their education had my students been a part of collective action or collaborative decision-making? According to them, never.

Because I believe in what we are doing, each semester when we hit roadblocks, I take a deep breath and let the process work. It is easy to take over and solve problems for students; what is difficult is letting them solve those problems themselves. I wait to be invited into their conversations; sometimes, they want to check in with me just to test the waters, whereas other times, they come to me as simply another resource person in the room. When they get stuck, it becomes my role to ask them questions to enhance and guide their thinking rather than to provide them with answers. While much of class time is given over to the group for their collaborative planning, we are also holding discussions over the readings, watching videos of democratic classrooms and inquiry-based teaching, and sharing children's literature that promotes democracy and inclusion of all. The projects that have come forth as a result of this process have been heartwarming, meaningful, and have made a difference not only in the lives of others but also in my students' lives.

The group collaboration projects, as well as the individual projects, have been as diverse as the students engaged in their development. At each juncture, students kept the integration of literacy, community, democracy, and social justice at the forefront of their thinking, planning, and decision-making. Each of the projects was developed using a moral compass. The students grew in their thinking, learned much about themselves and who they want to be as teachers, and developed agency and confidence that they could effect change by being leaders in their schools and communities. Here is a synopsis of the group collaboration projects developed over the past several years, in the words of the teachers who developed them:

> **Democratic Education in Action: A Professional Development Workshop for Future Teachers, Fall 2009:** Our rationale for this project is to inspire future teachers and to demonstrate ways they can empower their students and work for change in their classrooms. Dedicating ourselves to creating democratic classrooms and schools is one way that we can work to create a truly just society. We need to teach young people to think critically about social issues, and to ask questions about the inequalities that they see in the world. If we want our students to challenge injustice, we need to empower them and let them know that their voices deserve to be heard. The professional development workshop aims to make future teachers aware of the "democratic classroom." We hope to show teachers that by offering their students a voice, and a legitimate claim to their own education, they can be more productive, and more able to be advocates for themselves. We hope to increase the level of success that students are achieving in schools by allowing them to help build their classrooms, and their curriculums. Democracy in a classroom can lead kids to be more involved in their own lives, which hopefully escalates to a more global voice. The end result will be more involvement in their community and the world at large. (Group Collaboration Paper, Fall, 2009)

> **Change for Books: Book for Change, Spring 2010**: The Ronald McDonald House, whose mission is to create, find, and support programs that directly improve the health and well-being of children, is in constant need of books for its families. Our Literacy Education graduate level class is collaborating on a project which involves literacy and social action. We will be collecting money, as well as new books, for the Ronald McDonald House in order to help individuals of all ages escape, relax, and develop active imaginations while in a time of crisis and need. Participating in this project will help us and our communities become active citizens as well as enrich the lives of those in need

through powerful literary materials. Overall we raised $600 to spend on books across all age ranges. (Group Collaboration Paper, Spring, 2010)

Ending Child Hunger, Fall 2010: "... the first essential component of social justice is adequate food for all mankind. Food is the moral right of all those born into this world." (Norman Borlaug) As teachers, we know all too well the effects that poverty and hunger have on the North Country as well as on our students. We know the reality is that for some students, the only adequate meals they receive are at school. Knowing this, the thought of any child going all weekend or holiday vacation without any substance haunts us all as we see their withdrawn faces when we sit down at our own dinner tables. According to Tufts University Center on Hunger, Poverty and Nutrition, undernourishment is directly linked to delay in learning ability and brain development. The longer a child goes without receiving proper nutrition the greater the learning deficiency, this can also delay social growth as well. The good news is "as children become better nourished, they become better able to learn" (Knox, 2001). We recognize the limitation that hunger places on the learner and together we stand to say—No more! We care and want to do whatever we can to improve the health of children everywhere, starting with those living in our own community, children who need us to be their voice. Hopefully, through our efforts we can bring awareness about child hunger, which is a very real issue living right here in the North Country. We hope by supporting the Plattsburgh based Interfaith Food Shelf, we can improve the lives of some families, and in doing so we hope to change the world, one person at a time. (Group Collaboration Paper, Fall, 2010)

Chemo Literacy Activity Bags, Spring 2011: Literacy Education Graduate Students in EDR513: Literacy for Democratic and Global Societies collaborated on a class project to provide "Chemo" Literacy Activity Bags for children receiving cancer treatments at the Vermont Children's Hospital Fletcher Allen Pediatric Health Care unit. Through weeks of bottle drives in the local community we raised over $1,100.00, enough to purchase materials and create literacy bags filled with books, puzzles, games, dolls, markers, crayons, blankets, and stress balls we wanted to donate to the children. The drive was so successful we were able to create and fill 60 individual bags for children Birth through age 13 and, in addition, donate an additional $400.00 to the pediatric unit so materials could be replenished as needed. Books representing an array of developmentally appropriate reading levels and materials will provide opportunities for children to engage in a variety of literacy experiences while receiving their cancer treatments. Project requirements were broad and had to reflect elements of literacy, democracy, social justice, and community. Everyone had to be involved and have a voice in decision-making and problem solving. The environment was one that fostered a contagious, positive energy that helped the project progress at a good pace. None of us could have predicted the amount of money we ended up raising to benefit this cause. The support we received from the community was amazing. An outstanding number of individuals donated bottles and money which helped us go above and beyond our initial goal. This project was both a unique and memorable experience for each of us as individuals as well as education professionals. (Group Collaboration Paper, Spring, 2011)

Books for Babies, Fall 2011: As a class of individuals who pride themselves on improving our ability to create change and promote social justice, we collaboratively decided to promote literacy from a young age by starting right where life happens—the birth ward at Champlain Valley Physician's Hospital in our community of Plattsburgh. In order to create a just world, it is necessary to raise awareness, analyze that awareness, and then use the awareness and analysis to create and implement a plan of action for change. Our Books for Babies project was created in order to promote early literacy and provide new parents with the supplies and information needed to jumpstart their child's early literacy development. In addition to providing information on the importance of family literacy, our goal is to send new parents home with three books: one baby book, one toddler book, and one school-age book. With the $600 we raised, we were able to construct 75 book bundles to donate to the birth ward as well as 25 extra books to be used in the hospital waiting room. As teachers and avid readers, we know how important it is for children to be exposed to literacy at a young age. By donating books to the families of newborns we hope to help them see the importance of reading in their children's lives. (Group Collaboration Paper, Fall, 2011)

Hurricane Sandy Relief Effort, Fall 2012: Dr. Bobys' Literacy for Democratic and Global Societies class engaged in a large scale collaboration project to provide relief for those impacted by Hurricane Sandy

> on Long Island. The class traveled together to New York City to help clean up in the aftermath of the storm, worked together on a book drive to help replenish libraries that had lost literature collections, and had a winter clothing drive where clothes were donated and delivered to the MLK Center in New York City for those in need. The Literacy Education graduate students left for New York City at 3:00 AM on Saturday, December 1st. By 9:00 AM they were at the "All Hands Volunteer" organization where they were provided with t-shirts and directions to the site where they would be working. Everyone was surprised to see the devastation that occurred to the house they worked on, which had about 8ft of water damage as a result Hurricane Sandy. Because the water damage was so significant, the entire basement needed to be cleaned and gutted. Some of the jobs required on the site included tearing up the floor, knocking down walls and carrying the debris out to the curb. "Learning and reading about social justice is such a privilege but acting upon our learning, now that is a life lesson we will always remember," writes Marijane Stanley. Kristen Levis writes, "It took great communication, collaboration, teamwork, and effort to make this happen. As we were driving down to the city with cars full of donations for people in need, I could not help but think about how much of a difference our little class was making. I truly understand the importance in taking action and being a participant in communities everywhere." (Group Collaboration Paper, Fall, 2012)

Individual projects were just as compelling, but definitely personal and spoke more to the passions of each of the individuals. Too numerous to list, students have engaged in projects from creating books written in Mohawk to donate to a reservation school, to educating their third-graders on animal care while volunteering time at a no-kill animal shelter, to raising money for victims of Hurricane Sandy by having first-graders create pictures that were displayed and sold at a silent art auction during conferences at their elementary school. I can honestly say that the impact of these projects on my students has opened the door for them not only to bring democratic learning into their own classrooms but also to maintain their own enthusiasm for personal involvement as a way of life rather than an assignment for a class.

5.8 NEVER REALLY AN END, BUT WHAT DID WE LEARN?

> *As teachers in a democratic society...we must come to think of ourselves as ethical actors with a large degree of flexibility and autonomy in order to attend to and support the growth of children. In a democracy, teachers must be models of thoughtfulness and care, exemplars of problem solving and decision making; people capable of asking deep questions, drawing necessary connections, incorporating the surprising and the unexpected and the new as it occurs in classroom life—Bill Ayers*

At the end of each semester, I ask my students to reflect on their work during the course of the semester. I ask them to start by rereading the enduring understandings and essential questions. I ask them to go back and look at their thinking in the reading partnership reflections. It is their charge to think about how all of what they have accomplished during our semester together has impacted their thinking around those understandings and questions. They have been in partnerships, groups, and have worked as individuals. They have lived in a democratic classroom where their questions determine the direction the class takes. They have made a difference in the lives of many outside of our classroom. Although we have come to the end of our journey—for now—the idea is that this is only the beginning of the journey that will become their life's work. I have included a few excerpts of their learnings; their voices speak to how self-knowledge impacts the ability to teach ethically.

5.9 STUDENT VOICES

> **Ali:** What is a democratic classroom? Before taking this class, I thought that it was as simple as letting students take a vote on a class project or activity, but now I realize that it is so much more than that. It is about letting students have true choice in their learning. It is about engaging the students, and teaching them about the social responsibilities they have within their community and the impacts they can have on society. It involves exploration and identification. It means that the teachers and the students are equal with one another because in democratic classrooms, everyone is a teacher and everyone is a student. I want my students to be able to think critically about the world around them. I hope

that they will question the limitations placed not only on them but also the holds placed on others. I want them to work collaboratively with one another and respect one another's thoughts and ideas. I hope that they can embrace and enjoy the differences in the world, but I realize that to teach my students these ideas, I, myself must model these same principles. (Final Reflection, Fall, 2011)

Teri: Over the course of my journey through *Literacy for Democratic and Global Societies*, I have evolved in not only becoming a teacher of literacy, but in how to become a teacher of our world. I have learned so much about how to support and advocate for my students through change and the implementation of democracy within my classroom. Democratic classrooms are not just about voting; it is about collaboratively strategizing to resolve the issues around us larger than ourselves. This is how I wish to involve my students in real-life experiences. I strive to have a respect for each other as we work towards a common goal, take risks, while ensuring that the best interest is met for all students. I believe democracy should be at the heart of all of our classrooms to create a voice for each student within the classroom. We must embrace our students' individuality and perspectives as this understanding must drive our curriculum. I believe as we go through our lives, we are creating a book—a story of our most notable experiences. My book has many pages, containing several experiences, and I find that this class has filled a large portion of this book. I have grown significantly as an educator and an advocate for change.

Logan: I think that what my learning this semester comes down to, in simplest form, is that our students are "real people living real lives in the real world" (Beane, 2005, p. 71). In order to honor that truth, we must teach them democratically, empathetically, and constructively, all the while immersing them in meaningful literacies that will aid them in becoming active, critical thinking members of society who are indeed real people embarking on real lives. I think that our class discussions helped many of us, including myself, in the continuation of uncovering who we are as teachers and people. I would go as far as to say that our beliefs play a role in defining us, and our beliefs about teaching play a role in defining the teachers we will become and the students we will impact. For me, at their deepest and most intense, our discussions made me think about what I believe in and who I am at my core. What is scariest about realizing who you are at your core is that there is no turning back. I cannot possibly imagine voluntarily teaching in a manner that I do not agree with and know will be detrimental to my students' growth. I would feel like a fraud knowing what I know now, and not becoming an active agent in working to create change in our field. (Final Reflection, Fall, 2011)

Meg: I have learned how literacy, democracy, ethnicity, race, gender, and other social constructs have affected how I communicate in a diverse democratic society. In learning these things about myself and about my world, I now have a better understanding of myself and of my students in relation to promoting change, justice, and democracy. In order to make sense of the world I live in, I must become knowledgeable about the instances of change, justice, and democracy in my environment and beyond. I was guided through this course by using my prior knowledge to act as a vehicle to propel my new knowledge. In Frank Smith's book Reading Without Nonsense, he theorizes that "what we have in our head is our only resource for both making sense of the world and learning more about it" (Smith, 2006). When I participated in the reading, writing, speaking, and living that this course offered me, I used my own resources of the world to learn more about it. When I participated in all of these activities, I used my surroundings to make better sense of everything I was learning. Throughout the readings and all of the other components of class, I have learned who I am, what I believe in, and how I want to live my life and my classroom. Before this class, and more particularly, 3 years ago and beyond, I was too timid and shy of a person to speak up. Now, I have become a strong individual who is not afraid to speak up, and most importantly, not afraid to speak out. (Final Reflection, Fall, 2011)

Kewsi: The component of this class that really transformed my experience was the projects that we did around democracy, social justice, community, and literacy. These experiences helped us to live the life of democratic education, which is an experience that I will take with me into my classroom when my students do inquiry learning. The enthusiasm, the planning, the collaborations, the stumbling blocks will all allow me to empathize with my students. If I did not have the experience participating in such an elaborate process, I would do my students a disservice because from what I have learned, my experiences in school do affect the way I teach. If I introduce students to projects like the ones I participated in they will become critical thinkers, inquirers, active citizens, and problem solvers capable of making a difference in a democracy. I believe that we need to provide students this opportunity in school in order to prepare them for the real world. Students need to grow up learning to share their opinions, values, thoughts, and beliefs. More importantly, they need to know that their voices matter, so that they will become empowered to make change in the world. (Final Reflection, Fall, 2012)

Amanda: Where I took the most from this semester was our classroom discussions on our readings; the debates of ideas and opinions were refreshing. Books were selected to put you inside the heart of a classroom that is making all the ideals work. The authors were honest about the difficulties and bumps in the road; I felt like they were a part of our classroom discussions in a way. I was reminded that doing social justice or inquiry or making more room for democracy and choice do not make more work for the teacher, but they do make more room for students and creative minds. This class gave me strength not only to take risks in my classroom, but I feel like many teachers would benefit from being reminded of some of these ideals and philosophies and their applicability to the classroom. I can envision myself working towards a professional development series that put children at the true center and inquiry, democracy, social justice, and critical literacy as the means of reaching the standards. (Final Reflection, Fall, 2012)

Sarah: I learned a lot about myself as a teacher in this class, and I learned what type of teacher I want to be. I have learned that if I sit back and continue to allow practices that are not beneficial to our students I am just as guilty as the teacher facilitating them. I was intimidated to start my graduate courses, and I was even more intimidated when I walked into EDR513. I looked around the room and saw classmates that were older and more experienced than I was. I felt timid in the beginning, and I was nervous to speak my mind in the fear of sounding naive or even dumb. However, when I started to get my voice in the class, I noticed that it was not only listened to by everyone, but my ideas and comments were accepted and encouraged. I realized that the reason I became so open and communicative in the class was because I was in a comfortable and safe learning environment where democratic practices were being enforced. I did not have to worry about being criticized after speaking. Everyone's voice was heard. We all had a say in the curriculum for the semester, we had a say in what we would be doing for homework assignments and we had a professor that was open and willing to hear all of our comments and questions about the course. This is not the case in all courses, and it was both refreshing and useful to see the practices of democracy we are being taught in action in our own class. This class has changed me. It has made me realize my potential as a future teacher as well as the potential of my future students. This class has made me understand the importance of using literacy in a way that can help shape and form our students, classes and schools. It has made realize the effect that just one person can have on a class, a school, a community or even the world. It has made me eager to advocate for the rights of all students and teachers, and most importantly it has made me realize that I can make a difference. (Final Reflection, Spring, 2011)

Harmony: Everyone comes into the education profession for a different reason, but usually we want the same outcome for all of our students. It is common for educators to talk about preparing students to be informed citizens, capable of being active participants in our democratic system. It is much less common for them to help students understand why they should bother. After taking this class I feel like I am empowered with the tools to show students how they can become active participants in our democratic system. I came into this profession because I wanted my students to realize that they are important members of society and they can achieve and grow into whatever they desire as adults. This is such an important idea because as teacher we can help our students see how they can become active members of their community, how they can be a part of creating a democratic school and community, one that includes social justice. This experience reinforced ideas about accepting all students, treating students equally, giving choices, and meeting the individual needs of every child. These are the most important and crucial pieces when it comes to teaching democracy and social justice. (Final Reflection, Spring, 2011)

Joe: To recreate your own democratic way of life, you need to go off the beaten path, the road less traveled, to find a vast and magical world that is waiting to be explored and conquered. Yes, the view will be more scenic but that doesn't mean the drive will be an easy one. If anything it will be filled with plenty of twists and turns and plenty of bumps in the road. It will take many years to master and in many cases, it may never be mastered because the road to democracy is constantly changing. As you travel down that road you will pick up a plethora of ideas from fellow peers and colleagues. It is in your best interest to listen, and listen carefully for any ideas you may be able to utilize in your own democratic way of life. Listen for any directions that may help lead you in the right direction. Don't just stick to those well known stops along the way. Democracy, as I now know, is much more than just having a voice and making decisions as a group. It's about teaching your students to be proactive, taking control of a situation, making crucial changes and decisions regarding the community, the nation, the world. It's about taking the time to do the research, analyzing the problem/situation from all angles, working collaboratively and cooperatively to transform the problem/situation into a solution. As teachers, we need to teach our students the beauty of social justice and the power it can have on those involved. We need to engage them in realistic, critical thinking and problem-solving community action programs that get them thinking about the big picture and the impact they can have on others. Not only will it be beneficial in the short run but the long term benefits later on in their professional careers will make them successful, ethical, and responsible citizens who will more than likely carry these same beliefs and practices with them as long as they live. (Final Reflection, Spring, 2011)

5.10 IF NOT YOU, THEN WHO?

The personal right to dignity...includes the right to think for oneself, to be fully informed about important issues, to hold beliefs of one's choosing, to have a say in what and how things are done, to pursue personal aspirations and growth, to be free from oppression, and to experience just and equitable treatment—James Beane

In today's world of teacher accountability, high-stakes testing of children, federal, and state curriculum mandates, lack of funding for schools, and an imbalance and lack of access to teaching materials and resources, when do teachers have the opportunity, energy, or capacity to teach for equity and social justice? When are they able to practice their craft without interference from those furthest away from children? And how do they develop the agency and the voice to work against the status quo? These are the questions and concerns my students bring to our class discussions.

They are nervous and unsure of the challenge of creating this environment for their own students. My question to them, throughout the semester is always, "If not you, then who?"

As I read and reread my students' reflections of their experiences in *Literacy for Democratic and Global Societies,* and as I reflect on my own experiences as their professor, I am struck by our collective understanding that self-knowledge does, in fact, impact our ability to teach ethically. I believe my students have embraced the enduring understandings and essential questions that we have continually revisited throughout our work together, and I trust in their commitment to demonstrating their learning through living and acting on their beliefs.

Our time together is only the beginning of a life journey where we become empowered to impact the lives of others. With each experience, we take a step forward. This path is not an easy one, and my students discover their ideal may not be realized right away. However, with commitment and patience, and with constant reflection, I know my students will continue to work toward creating classrooms of love, dignity, hope, democracy, and social justice. If not them, then who?

KEYWORDS

- **Community building**
- **Decision-making**
- **Democracy**
- **Teaching**

REFERENCES

Ayers, W.; To Teach: The Journey of a Teacher. 3rd Ed. New York, NY: Teacher's College Press; **2010**.

Banks, J. A.; Diversity and Citizenship Education: Global Perspectives. San Francisco: Jossey-Bass, **2004,** 3–15.

Banks, J. A.; Teaching literacy for social justice and global citizenship. *Lang. Arts.* September, **2003,** *81(1),* 18–19.

Beane, J. A.; A Reason to Teach: Creating Classrooms of Dignity and Hope. Portsmouth, NH: Heinemann; **2005**.

Cherland, M. R.; and Harper, H.; Advocacy Research in Literacy Education: Seeking Higher Ground. Mahwah, NJ: Lawrence Erlbaum Associates; **2007**.

Dewey, J.; Democracy and Education: An Introduction to the Philosophy of Education. 1966 Edn. New York: Free Press; **1916**.

Dewey, J.; The Public and Its Problems. Chicago: Swallow; **1927/1954**.

Dewey, J.; How we think. A Restatement of the Relation of Reflective Thinking to the Educative Process. Revised Edn. Boston: D. C. Heath; **1933**.

Engel, B. and Martin, A. (eds.); *Holding values: What we mean by progressive education. Essays by members of the North Dakota study group.* Portsmouth, NH: Heinemann; **2005**.

Freire, P.; Pedagogy of the Oppressed. New York: Continuum Publishing Company; **1970**.

Furman, G. C., and Shields, C. M.; *How can educational leaders promote and support social justice and democratic community in schools?* Paper presented at the annual meeting of the American Educational Research Association, Chicago; **2003,** April

Greene, M.; Releasing the Imagination: Essays on Education, the Arts, and Social Change. San Francisco: Jossey-Bass; **1995**.

Johnson D. W.; and Johnson, R.; Cooperation and Competition: Theory and Research. Edina, MN: Interaction Book Company; **1989**.

Johnson, D. W.; Social interdependence: the interrelationships among theory, research, and practice. *Am. Psychol.* **2003,** *58(11),* 931–945.

Johnson, D. W.; Reaching Out: Interpersonal Effectiveness and Self-Actualization. 10th Ed. Boston: Allyn & Bacon; **2009**.

Knox, C.; Hunger and the ability to learn. Retrieved on November 23, 2010 from http://www.homefamily.net/index.php?/categories/foodnutrition/hunger_and_the_ability_to_learn/; **2001**

Smith, F..; *Reading without nonsense.* (4th ed). NY, NY: Teacher's College Press; **2006**

Wheatley, M.; Bringing schools back to life: schools as living systems. In: Creating Successful School Systems: Voices from the University, the Field, and the Community. Boston: Christopher-Gordon Publishers; (retrieved from http://www.margaretwheatley.com/articles/lifetoschools.html), **1999**.

CHAPTER 6

BREAKING THE MOLD: CHALLENGING PREVAILING MYTHS THROUGH CROSS-CULTURAL EXPERIENCES

CAROLINE KNIGHT and JAMIA THOMAS RICHMOND

CONTENTS

> *I think sometimes we assume that it's easier to work with children who are middle-class or above, and then the low-income students need a whole lot more attention—they are just a whole lot more needy. But the program in general has shown me that these children are just as intelligent, just as capable as those who have more advantage.*

6.1 TUTOR 6

Demographic trends in the United States project increasing diversity; by 2042, minorities will comprise the majority population of the United States. The number of White, non-Hispanics will decrease to 46 per cent of the total population by 2050, whereas ethnic minorities are projected to increase significantly (US Census, 2008). The District of Columbia, California, Texas, Hawaii, and New Mexico are already majority–minority entities (US Census, 2010). Ethnic minorities comprise 62 per cent of the nation's children, rising from 44 per cent in 2008 (US Census, 2008). In July 2011, for the first time in US history, minorities represented over half (50.4%) of the population of children under 1 year old (US Census, 2012). As this national trend continues, cross-cultural interactions will increase in frequency. These interactions can be positive or negative, dependent on the level of cultural competence held by the individuals involved.

Our individual identities influence the ways in which we perceive others and society perceives us. Nieto and Bode (2008) explain that our identities frame how we experience the world. They assert:

> Our society, among many others, categorizes people according to both visible and invisible traits, uses such classifications to assign behavioral and cognitive traits to these categories, and then applies policies and practices based on these categories that jeopardize some people and benefit others, causing particular damage to groups of people sharing specific identities. (p. 66–67)

To interact positively, leaving little to no damaging effects on others, one must have both cultural knowledge and sensitivity. Flaskerud (2007) described cultural knowledge as "actively learning about the community—its ethnicities, languages, origins, immigration or migration history, acculturation level, economy, sources of income, family and social structures and roles, value systems and beliefs, educational levels and literacy, geography, and ecologic environment" (p. 121–122). Cultural sensitivity can be defined as an awareness of one's own cultural beliefs and practices, being nonjudgmental of others who have differing views, and the ability to be flexible and adapt to cultural contexts and situations (Flaskerud, 2007). To be effective professionals in any field, college students must develop skills that demonstrate cultural knowledge and sensitivity; to develop cultural knowledge and sensitivity, they must first examine their own unconscious biases based on race and class.

Nieto and Bode (2008) identify social class as an example of the complex connections between difference and discrimination. They point out that despite the pervasive societal belief in social class mobility, classism prevails in the United States, where economic inequality is now greater than in many other wealthy nations. Society blames individuals of low economic status for their situation because people have accepted the myth of the culture of poverty, and considers poverty as a never-ending cycle. What Gorski (2008) describes as a *culture of classism* may lead even the most sincere professionals to pity children living in poverty and have low expectations for their achievement. He challenges us to question the prevailing myth about a culture of poverty and recognize that poor people do not necessarily share monolithic and predictable beliefs and behaviors. Gorski identifies four aspects of the myth: (1) poor people have an inadequate work ethic due to lack of motivation, (2) poor parents do not value education, (3) poor people are *linguistically deficient*, and (4) poor people tend to abuse drugs and alcohol. He writes, "The most destructive tool of the culture of classism is deficit theory. In education, we often talk about the deficit perspective—defining students by their weaknesses rather than by their strengths" (p. 34).

The growing diversity of our nation and the increasing frequency of cross-cultural interactions demand that preprofessionals prepare to work effectively with individuals from diverse backgrounds. In many colleges and universities, challenges remain in creating learning experiences that develop cultural competence that can be maintained once students enter professional practice (Angie, Davis-Maye, and Zugazaga, 2007). Many schools in the social services field (e.g., education, social work, and psychology) require foundational coursework in diversity. Most often, these stand-alone courses are limited to experiences through lectures, readings, multimedia, and guest speakers (Rone, 2008); but a gap persists between theory and actual practice of the skills (Angie, Davis-Maye, and Zugazaga, 2007; Keengwe, 2010). Increasingly, universities recognize the need for cross-cultural experiences to enhance the development of cultural competence. While diversity courses are necessary and beneficial to preprofessional students, immersion in cross-cultural experiences, coupled with opportunities for guided self-reflection, are essential for these students to "become knowledgeable about other cultures, reduce bias, develop respect skills, and become accepting of *others*" (Keengwe, 2010, p. 7).

Vaughan (2005) created an immersion experience to help graduate preservice teachers discover what it would be like to be part of a minority group. He maintains that preservice teachers must be adequately prepared to work with students from diverse cultural backgrounds. In our pluralistic society, he writes, "... we are sometimes socialized in ways that encourage us to harbor prejudices, stereotypes and misconceptions about individuals and groups that are different from ourselves" (p. 28). Vaughn's graduate students came away from their immersion experiences with genuine understanding about minority students in classrooms, and with motivation to learn more about cultures different from their own. Cross-cultural encounters give students authentic experiences that assist them in learning abstract concepts. Through experiences outside the college classroom, students benefit from opportunities to connect theory to practice (Chavan, 2011). Brock, Wallace, Herschback, Johnson, Warren and Poulson (2006) suggest that cross-cultural experiences give students opportunities to move beyond their comfort zones, explore unfamiliar cultures, and experience culture shock, giving them a deeper understanding of the displacement many culturally and linguistically diverse people may feel. Vaughan (2005) asserts that submersion into situations or environments outside of one's comfort zone are essential to understanding cultural diversity. To adequately prepare preprofessionals to work with diverse populations, institutions of higher education must move beyond standalone diversity courses and include cross-cultural experiences combined with guided reflection.

The purpose of this study was to document the impact of cross-cultural experiences on college students' preconceptions of families living in a low-income housing development. The authors interviewed 10 college students who participated as tutors with an afterschool literacy program in rural South Carolina for 11 weeks during the spring semester. The program took place in the community center of the housing development, and tutors had the opportunity to meet family members of the children they served.

This chapter focuses on the answers elicited that reveal the college students' preconceptions and beliefs about diverse learners and how those preconceptions changed over the course of a semester. Initially, the college students expected to encounter behaviors that would have mirrored the stereotypes identified in Gorski's *Myth of Poverty*: inadequate work ethic, poor parenting, linguistic deficiencies, and substance abuse. The researchers expected to find significant changes in the college students' perceptions about working with learners from culturally diverse backgrounds at the conclusion of the experience.

6.2 PROCEDURES

All the tutors attended a mid-sized liberal arts university in South Carolina. They were identified through participation in a rural afterschool literacy tutoring program. In a study describing the experiences of individuals with a certain concept or phenomenon, a recommended sample size ranges from 6 to 10 participants (Creswell, 1998). In this study, 10 college students participated in post-experience interviews. To protect the identity of the students, each individual was assigned a pseudonym corresponding to numerals 1–10 (See Table 6.1).

TABLE 6.1 Demographic characteristics of university tutors

Pseudonym	Race	Gender	Undergraduate major
Student 1	AA	F	Elementary
Student 2	C	M	Elementary
Student 3	C	F	Business
Student 4	C	F	Elementary
Student 5	C	F	Elementary
Student 6	AA	F	Psychology
Student 7	C	M	Psychology/sociology
Student 8	C	F	Psychology
Student 9	AA	F	Sociology
Student 10	C	F	Health promotion

Following the experience, the researchers each interviewed five students, recording the interviews for accuracy in reporting results. In addition, notes were written to clarify responses and pose additional questions to the students during the interviews. One student provided written responses to the interview questions as an alternative to participating in a recorded interview. This modification was made to accommodate the special needs of the student. A list of 11, open-ended questions was compiled for the post-experience interview. Three questions addressed the college students' preconceptions of the children and the community in which the experience took place. The remaining questions focused on how, if at all, their conceptions shifted during participation in the cross-cultural experiences. A list of the questions is found in Table 6.2.

TABLE 6.2 Tutor interview questions

1. What is your undergraduate major?
2. Why did you volunteer to be an afterschool literacy volunteer with freedom readers?
3. What did you hope to give? To gain?
4. What rewards have you experienced as a freedom readers volunteer tutor?
5. What were your biggest challenges?
6. Is this your first cross-cultural tutoring/teaching experience? What role did race and/or class play in your experience?
7. Has this experience changed the way you think about working with children in low-wealth communities? In what ways?
8. What did you find to be the strengths of your young scholar? Of his/her family?
9. What new ideas do you have regarding your career as a result of this experience?
10. What did you learn about yourself?
11. How have media images shaped their images of low-wealth neighborhoods? How did that change for them?

6.3 FINDINGS

The college students responded to several questions about their initial attitudes or expectations for working with children from a low-income community. The responses are organized according to the preconceptions they represent regarding the first three ideas that Gorski identified as the *Myth of Poverty*. Another category of responses show how the college students' preconceptions of the environment in a low-income community changed over time.

6.3.1 *PRECONCEPTION 1: INADEQUATE WORK ETHIC*

Some of the college students' answers indicated they had initially assumed that the children needed mentoring because of the poor families from which they came. Student 10 said, "I feel like the children come from disruptive homes ... we want to give them hope and we want to give them aspiration to become better readers, to become better students in school." Student 10 explained, "Dealing with children that perhaps have come from disruptive homes and you can tell that, I feel, by how they act. They wouldn't want to concentrate on the book they were reading."

Four college students mentioned that they initially wanted to serve as a "positive role model." Clearly, the assumption was that the children did not have role models who had an adequate work ethic. As Student 9 said, "I wanted to give, like, courage, because I know a lot of young kids don't have anyone to look up to, so I wanted ... I actually go there hoping to be a role model for whatever child that I have, and encourage them to always give it their all."

At the culmination of their cross-cultural experience, however, many of the college students reported that their perceptions had changed. When asked about the strengths of her child, Student 5 reported that the child "would challenge herself and push herself." She also described her scholar as reading with expression, using comprehension strategies, and being able to re-tell what she had read the week before. Student 6 described her child as being "very inquisitive. He loved to know why things worked the way they did," and he noticed many details.

Student 1 reported, "My scholar was persistent. She set a goal; she was determined to meet that goal." Student 2 said he learned that "I should not expect any less of someone because they are from a lower-class community."

6.3.2 *PRECONCEPTION 2: POOR PARENTING*

Several college students entered the tutoring experience expecting to see evidence of inadequate parenting. Student 9 expected that the neighborhood would consist primarily of single parents. One college student said that the child she tutored was "acting out a lot," which she initially attributed to his family. "I'm sure his mom didn't have time to sit down and focus with him on this one very important task, so that's gotten in the way of the progress." Another college student assumed that a child's low reading level indicated that she did not get much attention at home.

When asked, at the end of the semester, what they saw as the strengths of their tutee's family, the college students showed a noticeable change of perspective. Student 10 said, "You can tell he is very family-based, and he has a really tight family. His mom was really nice … she participated and interacted with (her son)." Student 9 said that her student's greatest strength was his family. "I think they are really connected to him and he's connected to them." She said she could tell they were a close-knit family because her scholar's younger brother was in the program, and he would never leave without him.

Two other children had siblings or cousins in the program, and their tutors noticed the family closeness. Student 1 said that her tutee and her brother talked everyday before the session started and used to share their snacks and check to see what books their sibling got. Student 5 mentioned that her tutee's cousins were in the program and "The cousins were very into their books, too, and into their reading." Student 6 said she could tell that her tutee's mother "really wanted him to be in the program."

Student 2 had positive things to say about his student's family:

> He had a great family system around him. He had a lot of siblings, two of them that were involved with [the program]. His older ones he really loved. When his brother got off the bus from high school, he would want to be out there every day, just to say "hey" to him. I've met his father, and he was a really nice man and his mother, and I'm sure they all support him really well and that's a great thing going for him.

6.3.3 PRECONCEPTION 3: LINGUISTIC DEFICIENCIES

"I thought they were all going to be struggling readers," one tutor told us. Several college students initially expected that the children would be "linguistically deficient" (Gorski, 2008), and linked that deficiency to their home life. Student 5 reflected on her preconception that the children would all be struggling readers and said, "Afterwards, I realized that they all came in there with a love for wanting to read and to learn and to help themselves be successful."

At the end of the experience, seven of the college students described strong literacy skills and/or said that their children were "really smart." Student 10 said, "He was such a good reader and he was really good at drawing." Student 9 was happy to report, "He was a great reader, and throughout the course he got better." Student 1 described strong personal reading skills: "My scholar had skills for phonemic awareness." She explained that the child could "sound out" words, and also understood what those words mean. Student 3 said that her child was very smart and "was always picking out ridiculously hard books."

The only college student who considered the role that schools might play in her child's achievement was Student 1, a nontraditional, African-American college student who was older than the other college students. She did not believe that the schools had encouraged the children to "seek their best" or to enjoy reading. She asserted:

> These children are literate—extremely literate—but they just don't… um … it's not the idea of the mainstream; their expertise is not in that arena, so then they are not given credit for what they know … it has to be impressed on their minds that they are literate, but they just have to come to place where they can function as the mainstream functions if they want to be successful.

Student 2 described what he learned when he said,

> He definitely exceeded my expectations when it came to how good of a reader he already was before I arrived; and, um, so I think that was good to sort of break the mold for me, so I do not have these kind of preconceived notions before going into it without knowing anything about the person.

6.3.4 PRECONCEPTION 4: DYSFUNCTIONAL COMMUNITIES

Student 2 admitted, "I guess I can honestly say that I didn't have the highest expectations going into it." Two of the college students felt that the housing development was very "sad"; and one of them was "shocked" by the small size of the two and three bedroom duplexes where most of the children lived.

Three college students eventually had positive things to say about the community in which they tutored. Student 2 said, "Some of the images from media or whatever are of violence and gang activity. But it was a great community. Really a family environment. All the kids are really nice and fun kids to be around." Another college student said that the community was really strong and that everyone seemed to know each other. "I thought that was really cool."

Student 5 noticed what she called "different family life styles." She said that in her own community she does not see children playing outside much anymore. By contrast, when she drove away from this housing development every week she saw children playing on the playground. She concluded thus: "They have more trust and know that people are going to be there in that community for them."

6.4 CONCLUSION

The responses of the students illustrate the critical need for pre-professionals to have cross-cultural experiences while in college. The students initially approached the children with faulty preconceptions based in deficit theory. They assumed that the children would be deficient in language skills, impulse control, and work ethic, and attributed those deficiencies to families living in poverty. Their perceptions of the children's abilities showed marked change by the semester's end, with most students becoming very enthusiastic about the children's efforts and achievements. Several students changed their opinions about the children's families, even identifying family closeness as their child's greatest strength. Three college students described the community itself in very positive terms.

Cross-cultural experiences alone, without guided reflection, are not sufficient to counteract prevailing societal myths. Implications for further study include increasing opportunities for reading and reflection during cross-cultural experiences, following the growth of preprofessionals who have repeated cross-cultural experiences over time, and increasing opportunities for college students to interact with community members. Further studies would give greater insight into the value of deeper, more reflective experiences.

Researchers could also assess the benefits of a variety of university courses for developing cultural competency. Diversity courses can provide important exposure to the vocabulary and concepts of fairness, equity and cultural competence. Other courses in human geography, anthropology, world religions and sociology can build a strong foundation for understanding of self and society. History courses can increase understanding of privilege, labor relations and civil rights. Global languages are essential for a well-rounded education, and lead to greater empathy for English language learners. Studying institutions where the faculty has made a concerted effort to foster cultural competence across the board may lead to improvements in general education courses.

The results of this study support Gorski's contention that the myth of a culture of poverty must be confronted and challenged. The myth leads people to assume that poor people do not value hard work or education and that they have linguistic deficiencies. Cross-cultural experiences, combined with guided reflection, help preprofessionals refute these stereotypes and develop deeper and more authentic understanding.

One college student summarized his new understanding about the children: "A lot of them are very bright, very positive. I never would have expected it. That was a totally wrong way to think."

KEYWORDS

- **Cross-cultural experiences**
- **Diversity**
- **Literacy experiences**
- **Tutoring**
- **University experiences**

REFERENCES

Angie, C.; Davis-Maye, D.; and Zugazaga, C. B.; Can cultural competence be taught? Evaluating the impact of the SOAP model. *J. Soc. Work Educ.* **2007,** *43,* 223–241.

Brock, C.; et al. Negotiating displacement spaces: exploring teachers' stories about learning and diversity. *Curriculum Inq.* **2006,** *36,* 35–62. doi: 10.1111/j.1467-873X.2005.00345.x

Chavan, M.; Higher education students' attitudes towards experiential learning in international business. *J. Teach. Int. Bus.* **2011,** *22,* 126–143. doi: 10.1080/08975930.2011.615677

Creswell, J.; Qualitative Inquiry. Thousand Oaks, CA: Sage Publications; **1998**.

Flaskerud, J. H.; Cultural competence: what is it? *Iss. Ment. Health Nursing.* **2007,** *28 (1),* 121–123. doi: 10.1080/01612840600998154

Gorski, P.; The myth of the "Culture of Poverty". *Educ. Leadership.* **2008,** *65(7),* 32–36.

Keengwe, J.; Fostering cross-cultural competence in pre-service teachers through multicultural education experiences. *Early Child. Educ. J.* **2010,** *38(3),* 197–204. doi: 10.1007/s10643-010-0401-5

Nieto, S.; and Bode, P.; Affirming Diversity: The Sociopolitical Context of Multicultural Education. 5th Ed. Boston, MA: Pearson**2008**.

Rone, T. R.; Culture from the outside in and the inside out: experiential education and the continuum of theory, practice, and policy. *Coll. Teach.* **2008,** *56(4),* 237–246.

US Census Bureau. An older and more diverse nations by midcentury. Retrieved December 11, 2013 from: http://www.census.gov/newsroom/releases/archives/population/cb08-123.html, **2008**.

US Census Bureau. Most children younger than age 1 are minorities, Census Bureau Reports. Retrieved December 11, 2013 from: http://www.census.gov/newsroom/releases/archives/population/cb12-90.html, **2012**.

Vaughan, W.; Educating for diversity, social responsibility and action: pre-service teachers engage in immersion experiences. *J. Cult. Divers.* 2005, *12(1),* 26–30.

CHAPTER 7

THE LEASTWISE OF THE LAND: TEACHING FOR SOCIAL JUSTICE IN A HOMOGENOUS ZONE

DAVE IASEVOLI

CONTENTS

> To give up on integration, while aware of its benefits, requires us to consciously and deliberately accept segregation ... Segregation, rarely discussed, scarcely even acknowledged by elected officials and school leaders ... is incompatible with the healthy functioning of a multiracial society—Orfield & Eaton, cited in Kozol, The Shame of the Nation

7.1 OPENINGS

What does the term "social justice" signify to new teachers today, in our larger, "multiracial society"? I wished to find out more about this after interviewing candidates for our M.S.Ed. program in a small institution in Upstate New York, and as I taught a variety of graduate courses in education. Our teacher education unit has created a "guiding philosophy" that posits three "claims"; the third includes this line: "our graduates are committed to being *multicultural competent educators* and teaching all students in their classrooms" (TEAC Brief, 2013; emphasis added). Thus, when I wrote interview questions for applicants to our new "Teaching and Learning Master's Program," I included this: *"If you purport to 'make a difference,' what does this mean?"* In our interviews, a disquieting phenomenon grew more and more apparent. My colleagues and I were disconcerted that not one applicant ever mentioned her work with poor or otherwise disadvantaged children. No one addressed the difficult topics of race, gender, sexual orientation, and children's perceptions of these differences. Given that all applicants were White—and that *all* of the students and teachers I have seen over the past 6 years identified themselves as White—did this phenomenon signal that such issues *did not matter* in Upstate New York? Further, I wondered what the real intention lay behind our program's mission to teach "multicultural competent educators": even if future teachers did indeed "teach all students in their classrooms," how could this possibly address issues of difference if *all* the students are White? Perhaps such teachers do not think that racial matters *matter* to them and the students they teach.

Part of us always wants to hold at bay such a disturbing suspicion, perhaps by considering such mitigating factors as the applicants' youth and lack of experience, or the actual field in which we worked and lived. That is, our geographic area remains predominantly White, with Black and Latino and Asian demographics all significantly below the national averages.[1] Nonetheless, all of our teacher-education programs espouse a "commitment" to multicultural education. I wondered if this is even possible when teachers, who are themselves White here, rarely work with students of color. I wanted to learn about these new teachers' understanding of "multiculturalism" and their intentions to bring into their own classrooms a curriculum that at least occasionally points to problems of inequality and injustice that beset our society. As Gwendolyn Brooks (2000) asks, in her poem "The Children of the Poor," "What shall I give my children? who are poor?/who are adjudged the leastwise of the land?"

7.2 OTHER VOICES

Kliebard's, *The Struggle for the American Curriculum: 1893–1958* (2004) outlines ideologies that have shaped the designs for what is taught in our public schools. The third development, from the 1920s, is the ideology he calls "social meliorism": "deriving its central thrust from the undercurrent of discontent with the American economic and social system ... [it] saw the curriculum as the vehicle by which social injustice would be redressed and the evils of capitalism corrected" (p. 154). Here, Kliebard describes the broader economic, political, and literary movements, such as the muckrakers and the novels of Upton Sinclair, that led to critics' descriptions of public education as serving the few—an elite population, either by birth or by talent. Further, Kliebard cites Dewey (in the latter's description of the presidential candidate Herbert Hoover) in 1928, as our greatest homegrown philosopher and deeply critical of our culture of wealth:

> [His] whole creed of complaisant capitalistic individualism and the right and duty of economic success commits him to the continuation of that hypocritical religion of "prosperity" which is, in my judgment, the greatest force that exists at present in maintaining the unrealities of our social tone and temper. (p. 151)

Kliebard's history concludes with the passage of the National Defense Education Act of 1958 (p. x); thus, he stops shy of entering the turbulent waters of the Civil Rights Movement and the effects of federal legislation upon a curriculum for social justice.

[1]According to the 2012 Census figures, in New York State's Warren County, African Americans comprise 1.1% of our population; Latinos 1.9%, and Asian Americans .8%. These figures contrast to 17.5%, 18.2%, and 8%, respectively.

Giroux's work (1981; 1991) explicitly addresses the role of education in the reproduction of dominant forms of culture—to the detriment of multiculturalism and social equity. As a neo-Marxist, he argues, in *Ideology, Culture, and the Process of Schooling* (1981), that schools perpetuate social injustices through the "guise of objectivity, fairness and merit. … the knowledge conveyed by schools reflects the principles of the dominant group … to legitimatize its power and enhance its capital accumulation" (p. 17). Giroux's work is prescriptive: He calls for a critical pedagogy in which teachers embrace their roles as "cultural workers" but not as transmitters; instead, they need to work as "transformative intellectuals … who help students understand how curricular knowledge may serve to liberate themselves to become critical and responsible members of a democracy" (1991; p. 92). Thus, in Giroux's classroom, teachers and students engage in a discourse that continually challenges both institutions and perceptions of the status quo. Students must become "media literate" to "breathe life into the notion of democracy by stressing a notion of lived community that is not at odds with the principle of justice" ("Border Pedagogy as Postmodern Resistance," 1991; p. 21). Giroux argues against any easy conclusions that we may make about race, racism, and racial differences in this country: "an anti-racist pedagogy cannot treat ideologies as simply individual expressions of feeling, but as historical, cultural, and social practices …" (p. 23).

There is an echo, sometimes, when reading neo-Marxists such as Giroux, of a wonderful TV ad (for Verizon) of some years ago. Here, a powerful CEO defends his choice of a cheaper and inferior phone plan, by announcing to his cowering subordinate that this is his way of "Sticking it to the Man!" The underling looks confused for a moment, then sheepishly notes "But, you *are* the man." Indeed, some of the strongest voices that cry out for social justice in our schools come from the Academe—a bastion of mostly White privilege. Thus, Lisa Delpit's work (2006; 2012) can read as both invigorating—and, to some, infuriating. To note that her first collection of essays, *Other People's Children: Cultural Conflict in the Classroom* (2006), caused a stir in education classes when it first appeared would be a gross understatement. Statements such as "… the 'open-classroom movement,' despite its progressive intentions, faded in large part because it was not able to come to terms with the concerns of poor and minority communities" drove many progressive educators to distraction. When I assigned this text in education courses in the North Country, new teachers reacted vehemently, for example, several said they wanted to burn their copies. "They complained about Delpit's 'one-note' tirade against White teachers who do not know how to teach Black children" (Iasevoli, 2009). Delpit's many critics in the academe pointed to her apparent lack of "scholarship" in her essays and speeches, but lost in the scuffle was a central point that rests upon a well-documented and rather depressing statistic: "… the turn of the [Twenty-first] century will find up to 40 per cent nonwhite children in American classrooms … 'Most teachers who teach today's children are white; tomorrow's teaching force will be even more so' " (p. 105). This last phrase can now be read as especially prophetic, as the "charter schools movement" has seen a majority of White teachers enter the teaching profession in mostly urban centers, where their students are mostly Black and Latino.[2]

Delpit seems to agree with Giroux's principle that demands a discourse of self-interrogation:

> We all interpret behaviors, information, and situations through our own cultural lenses; these lenses operate involuntarily, below the level of conscious awareness, making it seem that our own view is simply "the way it is." Learning to interpret across cultures demands reflecting on our own experiences, analyzing our own culture, examining and comparing varying perspectives. We must consciously and voluntarily make our cultural lenses apparent. (p. 151)

She explicitly addresses the need for White teachers to transform their understandings of what teaching *other people's* children demands. Yet in her latter work (2012), Delpit sounds more pessimistic about the possibility of equity in schooling, and retreats, after a fashion, into a more isolationist policy: *Multiplication Is for White People* provides many narratives of Black teachers working with Black children.

Jonathan Kozol's work has grown progressively more pointed in his attacks on the segregation of American public schools. Indeed, his greatest work may be *The Shame of the Nation* (2005), with its nearly despairing subtitle, *The Restoration of Apartheid Schooling in America.* Unlike Delpit, Kozol still pushes vehemently for integration in our schools. Here he cites Jack White, from an angry column in *Time* magazine (1995):

---

[2]In the "Knowledge Is Power Program," or KIPP schools, 95% of its graduates are Black or Latino. Of its 3000 teachers nation-wide, 38% of them identify as either Black or Latino (http://www.kipp.org/about-kipp/faq).

> Before we gave up on integration, we should have tried it ... Revive the civil rights movement, which went into limbo long before some of its most important goals were accomplished. ... The genteel race mixing that goes on among the elite is not substitute for a determined national effort to include poor nonwhite children in America's bounty .... (p. 216)

He argues, further, that the school "choice" and charter-school movements tend to further segregation of our public schools. The KIPP schools, however successful in raising students' standardized test scores, do not serve the same populations as their local public schools. Such divisions perpetuate what he labels "savage inequalities" that, in turn, result in the failure of some 70 per cent of Black males in New York City and Chicago to graduate (p. 282). According to the Schott Report (2010), "the graduation rate for Black male students for the nation as a whole in 2007/2008 was only 47 per cent"; the graduation rate for Black males in New York State as a whole is a depressing 25 per cent. That is, we ranked dead last in the nation. In exactly 1 year, educators and politicians will celebrate the sixtieth anniversary of the *Brown v. Board of Education* decision, but Kozol notes that *Plessy v. Ferguson* now appears the law of the land when you look at public school demographics. Here he cites education analyst Richard Rothstein:

> Even many black leaders ... have given up on integration .... They argue "a black child does not need white classmates in order to learn." So education policies ... "now aim to raise scores in [the] schools that black children attend. ... That effort will be flawed even if it succeeds." The 1954 decision ... "was not about raising scores" for children of minorities "but about giving black children access to majority culture, so they could negotiate it more confidently .... (p. 229)

Of course, Kozol and many of the other champions in the fight against social inequities in public schools focus on urban areas. What effects do their arguments have upon teachers in rural areas of the United States? A search for scholarly studies about "racial differences," "diversity studies," and "multicultural education" in rural areas of the United States yields scant results. One such study (Ayalon, 2004), concludes that "rural white residents tended to view education about minority groups as *irrelevant* to their communities" (p. 28, emphasis added). This observation resonated with me a few years ago, when a Master's candidate submitted one of his required projects, a teaching portfolio organized around nine themes. The final theme, of "Multiculturalism and Diversity," challenged students, but Wayne,[3] a high-school math teacher, approached this theme in a way that struck me as unique—and very disheartening. He wrote about math students' age-old difficulties with word problems, and he noted that test designers now routinely inserted names into such problems that students in his small town found quite unfamiliar, for example, Shaneequa. Wayne instructed his students to simply *change the names* into more familiar names, "so that they would not be distracted by strange names." He argued that this reflected his understanding of teaching about diversity.

7.3 CRIMINAL ACTS

If you wanted to reduce crime, you could, if that were your sole purpose, you could abort every black baby in this country and your crime rate would go down—William Bennett, Morning in America, 2005

If you scratch the surface of even the most politically correct elementary or high school teacher, would he nod—however ruefully—at the former Secretary of Education's hateful pronunciamento? In this Upstate New York region—worlds away from possible hotbeds of activism in large cities—most students and their teachers are White. An ethical question arises: How can integration matter in a monoculture? Since I teach a course titled "Exploring Multicultural Competencies and Literacy Instruction" each spring semester, I structured my investigation into the following questions with its students as collaborators:

In what ways does the stated goal of "teaching for social justice" impact new teachers in their education programs? Subsidiary questions include the following: (1) In what ways are teachers and teacher-candidates personally invested in this goal? (2) How do their definitions of "diversity" begin to change as they move through our programs? (3) What particular difficulties of teaching about change exist in our Upstate New York environment—a rather homogeneously White world? Since this course is required for all pre-service teacher-candidates and for those provisionally certified teachers who have chosen to concentrate on literacy,

[3]All names have been altered.

we read and wrote a great deal. There were four texts required; two of these deal explicitly with both issues and students that lay at least partially outside of the experiences of students in this course. The first, *Readings for Diversity and Social Justice* (Adams, Blumenfeld, Castañeda, Hackman, Peters, Zúñiga; 2010), is designed to raise awareness about issues of diversity and oppression—and to call for action to promote change. The second, *Killing the Sky: Oral Histories from Horizon Academy, Rikers Island*, contains six pieces from incarcerated students I taught back in 2004–2005.

The qualitative data that follow come from classroom discussions about these texts and select video clips, as well as the students' classroom writings and more mediated reflections in their "double-entry journals."

Out of the One...

In our very first session, I asked the students to note an "attempt at a definition" of "multiculturalism." Their responses ranged from the very straightforward and literal to the suggestion that there is a political aspect to the term. The majority of students focused on "differences":

Part of different areas of the world; ethnicities; difference/similarities; joining many aspects together; being "part" of something (e.g. nation); group, country, town; various ways of living; different foods, music, books, languages. Many—There is a wide variety of cultures, subcultures, etc., within any one group (e.g. schoolchildren). May include geographical, socio-economic, or ethnic differences, among others; "melting pot"

Two students seemed to infer that there is a certain degree of controversy embedded in this

"-ism":

It is implied that one culture is not superior or better than another, just different in some ways. The identification and acceptance of the diverse lives of people.

These last two "working definitions" proved useful for the whole duration of the course, and I told the students that they were so. I suspected that these definitions would prove to be *necessary* but not *sufficient* for the kind of work that is entailed by our claim that "our graduates are committed to being multicultural competent educators."

In our second session, the students received their own class' comments as notes, and I projected the reverse of a $1 bill onto the screen, magnified to show the bald eagle grasping the banner that spells out the motto: "*E pluribus unum.*" Only two could translate "Out of the many, One," and only one student applied this statement to the formation of One United States out of the original thirteen colonies. I suggested then that a belief in Multiculturalism inverts the order of the terms: "Out of the one, Many." In other words, if you truly espoused multicultural values, and purported to teach according to them, then the "plurality" or "many" or, simply, "differences" between people become more important than their unification or belonging to the same group or nation. And so, if teachers acted upon the belief that "one culture is not superior or better than another," then they must address the hard facts that discrimination persists as our dominant, White culture is invested in certain oppressive practices. Our primary textbook addressed such discriminatory practices (and not merely in the United States, but worldwide), in sections devoted to "racism," "classism," "religious oppression," "sexism," "heterosexism," "transgender oppression," "ableism," "ageism and adultism."

Other-ing

The students in this course reacted with some heat after their first assigned reading, "The Social Construction of Difference" (Johnson, 2010, in Adams et al.). In this chapter, the author argues that "To have privilege is to be allowed to move through your life without being marked in ways that identify you as an outsider, as exceptional or 'other' to be excluded ..." (p. 20). Students responded, in double-entry journals:

> I am so tired of reading stuff that tries to make me feel guilty just because I am White! I know plenty of White people who suffer, who are "oppressed" ... Johnson seems to suggest that all White people are privileged? What about the "99%"? that includes mostly White people. This is bull**** [sic]. You can be White and an outsider in many ways. I'm an outsider to most of my family just because I finished college ....

These reactions did not surprise me; instead, I truly believed that these students might have recalled an at least mild form of attack from the "politically correct police" they first encountered as undergraduate students. I pointed out that the author emphasizes the condition that "[W]hites can experience oppression for many reasons, *but not because they are [W]hite*" (p. 21, emphasis added), and asked the students to imagine ways that this understanding can shape classroom curriculum.

Here is where the matter grew much messier. There was a consensus within this class that all of the districts in which they worked or hoped to work—and those in which they had attended school themselves—were areas in which these issues *did not matter*. When I raised questions of race, racism, and racial inequities, they all quickly expressed their agreement that we, in the United States, did *not* yet live in a "post-racial" society. That is, they believed, they said that there still exists a great deal of racially founded inequity. More significantly, though, they insisted that they did *not* see themselves as "invested" in the goal of "teaching for social justice" simply because, in the local realm, they did not witness examples of *in* justice. One student, a woman who has already taught for many years, bid us recall that "All politics is local"; I responded that *de facto* no one can argue against this proviso. Thus, one voice after another spoke about progress in her or his immediate life, whereby a parent who used to spout racial epithets had stopped, or a school that used to be exclusively White now showed a number of biracial children. "It does not matter anymore in Williamstown that there are several Black families living here," said Terry. Several students agreed that elements of hip-hop culture had become the norm for many kids in the North Country; they cited this as evidence of a kind of multiculturalism. For instance, one teacher of high school students, who serves as the girls' varsity soccer coach, volunteered the information that the players started calling each other "niggaz" last year; none of them, of course, are Black. When he called *them* on the use of this term, they responded that it was not racist—they used it "in fun," and, further, it actually helped them cohere as a team. (The team enjoyed a winning record.) A majority of the students in this class agreed that the "n-word," and its many variations had lost much of its past power to offend; several students demurred by noting that only Black groups could use these terms without recrimination. Experience shows us that racial and ethnic slurs may no longer oppress members of a group they first demeaned, but may remain both hateful and harmful to others. For example, I may joke with residents of my old neighborhood in Brooklyn about who is the "Big Guinea" now, even as the term "guinea" was equivalent to the n-word for my father, who was born in Southern Italy.

This oft-discussed phenomenon of an oppressed group's "appropriation of the language of oppression" led me the following exercise. First, I showed several video clips: two from director Spike Lee, *Do the Right Thing* (1990) and *25th Hour*(2002); and a relatively recent clip from Quentin Tarrantino's *Inglourius Basterds* (2009). In both of Lee's films, we see a variety of characters viciously spitting out racial and ethnic slurs. Thus, in the first movie, for example, an Irish policeman in New York City lambastes Puerto Ricans. In the second, the protagonist, played by Edward

Norton, speaks into a bathroom mirror on which someone has written "Fuck You," and takes on *every* racial, ethnic, religious, and sexual group in the City. In Tarrantino's World War II movie, a German SS Officer invites a table of spies to play a guessing game: "The object of the game is to write the name of a famous person on your card .... When you finish writing, place the card face-down on the table, and move it to the person on your right. The person to your left moves his card to you. You pick up the card without looking at it, lick the back, and stick it on your forehead." In his example, he asks a series of questions to determine whose name is affixed to his own head—in this case, King Kong. (Significantly, the German officer's questions—and his fellow players' yes/no answers—can all apply to the African Diaspora, e.g., "When I went from the jungle to America, did I go by boat? ...Did I go against my will?...On this boat ride, was I in chains? When I arrived in America, was I displayed in chains?)

In our class' sixth meeting, we began by writing down on Post-It notes any stereotypical descriptors that a person of color might use to describe White people who live in rural environments. The examples of "redneck" and "hillbilly" came up immediately. For a few minutes, we wrote down nouns and adjectives and phrases that insulted—we "appropriated" such negative and even hateful terms as "KKK member" or "sex with sheep" or "incestuous"—and stuck them on each other's foreheads. Then a few volunteers tried to guess what was written about them.

The students reacted strongly to this game. They said that they found it very difficult to think like another and form questions to help reveal the printed stereotype. One student noted "I just never think of *myself* as a member of the KKK or like someone who loves NASCAR!" "I guess that's the point," opined a peer. "There must be a lot of Black people who never think of themselves as 'on welfare' or 'stealing watermelons' or some other crazy stuff." I closed this class session by asking the students to write for a few minutes about their students: what would the latter say if they were the other, if a lone White student were in a classroom of blacks?

Again, I encouraged these relatively new and inexperienced teachers to think of ways that this exercise with racial stereotypes could color their curricula. This was, granted, especially difficult for the two students who majored in chemistry as undergraduates. We agreed that to merely create an assignment for high-school students to conduct research into "chemists of color" would be worse than a token gesture. Such assignments tend to trivialize racial matters—assign them to the dustbin of irrelevance. Instead, we looked at several examples of news items from the front pages of *The New York Times* that offered possibilities for discovery in a variety of classrooms. Just a random sampling from one Sunday newspaper yielded stories about "Fears of National ID with Immigration Bill" and "Protecting Chimpanzees" (*The New York Times;* 2013). The students in this multiculturalism course then took a half-hour to brainstorm possible extensions into various content areas. Some of the topics that emerged were the basis for genetic studies; animal rights versus civil rights; human evolution in sub-Saharan Africa; voting demographics for the past two presidential elections, with a focus on the number of North Country residents who voted for Obama; figures from the great European waves of immigration versus the slave trade.

7.4 TOUGHER QUESTIONS

> *It takes courage to interrogate yourself. It takes courage to look at the mirror and see past your reflection to who you really are when you take off the mask, when you're not performing the same old routines and social roles. It takes courage to ask—how did I become so well adjusted to injustice?*—Cornel West, cited in Adams et al. p, 587

As the semester drew to a close, the students in this so-nicknamed "diversity course" reached their own tentative conclusions to the questions I posed (above). First, they all claimed that they supported diversity, and that they believed it a good in itself. As one student wrote,

"It is no longer acceptable to simply talk about injustice or social inequality, but rather we must now 'walk the walk' if we indeed hope to 'talk the talk.' " Another noted that she, at age 24 and a new mother, was beginning to see herself as old (!), and "in the eyes of our students, we are becoming 'OG's' like the men in *Killing the Sky* say. We should use our time as role models to initiate conversations about social justice and force them to begin thinking about the world outside of their own high school." All of the students agreed that their readings of the oral histories from Rikers Island inmates shocked—that they had never quite imagined experiences such as these just some 4 h south of our college. And so, with reference to a second question, about their changing definitions of diversity, these students reached consensus that it must stand as a relativistic term—and that their past and future (and our present) classrooms were *not diverse*. They were monoculturally White, and this creates and exposes limitations.

"Experience in everything is crucial," wrote one student in a final reading log entry; "if we do not have experience with another culture and/or race, how can we possibly hope to connect or to understand?" Others noted that they lacked courage in confronting the inequities described in our *Social Justice* textbook. "I ignored them," wrote Jon, "Most of them do not affect me in any way, so they are easy to become indifferent to …" Therefore, what else happens to teachers' critical consciousness of difference in institutions that show very little, if any, racial or even religious diversity? Immediately, in this class, students pointed to the fact that some kind of "diversity" exists within any group. Thus, in a homogenously white environment, teachers may uncover and examine differences in ethnic backgrounds. In addition, if they are somewhat braver, they might delve into the explicit differences in class backgrounds—and wealth. I reminded these teachers that they might also pay more careful attention to the possibilities that they treat male and female students in different ways; these behavioral patterns often reveal our acculturated biases. In short, there are many options available for teachers to "teach for social justice," but here I saw that there was no *ethical imperative for them to do so.*

For our final reading, I assigned an ancillary text, an essay that I had written some years ago that focused on another education class: "'A World of White and Snowy Scents': Teaching Whiteness" (2009), deals with racism—some subtle, some not—in the North Country. I expected that this article would raise eyebrows, and I was not disappointed. The woman who led the class discussion around my essay began by pointing out that "most people here are not like [those I described]. My parents never used the n-word and called people who did 'ignorant.' " Others chimed in that the main problems with my neighbors and fellow-firefighters (who wanted to learn to make hangman's nooses in one of our class sessions about knot-tying) were a lack of education and poverty. "We are not like that," said another student, with finality. Then, I quietly pointed out that perhaps they had read my article as *students*—instead of reading it as *teachers*. I suggested that although they were not like some of the people I described, some of *their students, in their own classrooms*, could indeed resemble my neighbor who anticipated that President Obama would be shot and killed. In other words, I needed, I thought, to make a final point about their own *responsibility* as teachers to work as "transformative intellectuals" (Giroux, 1991).

7.5 FURTHER IMPLICATIONS

> *What shall I give my children? who are poor, Who are adjudged the leastwise of the land* Gwendolyn Brooks, "the children of the poor, II"

Now I wonder if our graduate programs actually do a disservice to new teachers with this claim: "our graduates are committed to being multicultural competent educators and teaching all students in their classrooms" (TEAC Brief, 2013). In a way, the claim suggests a hypothetical situation, in which our graduates *May one day* teach a classroom integrated with Black and White students. But does this claim also suggest that there is an equivalence between "teaching all White students in their classrooms" and "being multicultural competent"? This, to me, is *not* ethical. It ignores the content that White teachers can and should bring into their all-White classrooms. If teachers and their schools, graduate students and their education programs, are truly committed to "teaching diversity," then all need to act upon this conviction in a much greater degree. To teach an inclusive classroom of individuals that includes students with disabilities is a *start*. To point to the significance that Black students are four times more likely to bear the label of a learning disability opens up a vast field of further learning opportunities.[4]

To *claim* that graduate students are "multicultural competent educators" because they have read and critiqued articles that address social inequity is a good start, perhaps. But to run an education program that purports to *change society* so that inequities diminish entails more active change on the part of both students and instructors.

Off course, the obvious question arises now: "Well, what can you do when there are no Black people in this area?" I suggest a number of beginning steps here. First, our programs could require a field placement in an

---

[4]"Statistics from the US Department of Education, Office for Special Education Programs for Fall 2008 show that although African Americans represented just 15% of all students, they represented 21% of students in the special education category of specific learning disabilities, 29% in the category of emotional disturbance, and 31% of the category of mental retardation. …Historically, special education has too often been a place to segregate minorities and students with disabilities." See http://aacld.org/theproblem.html

urban school in which Black children are in the majority—indeed, where the teacher might herself be the "Other." Second, in our curriculum courses, we could create assignments in which a teacher develops curriculum with another classroom of truly diverse students—a "pen-pal" classroom in a city—by means of Smart technologies and Skype.

Third, and most simply, open up our rural White classrooms: we could quite easily develop networks of "sister schools" that scheduled regular field-trips to visit each other. As our programs stand now, we instructors may be guilty of what Giroux describes as the "guise of objectivity, fairness, and merit" (1981), for the plain reason that we lack experience with Black students—or any students of color—in our monochrome environment. "Teaching multiculturally" becomes for us, at best, what Kant considered a hypothetical imperative: There is no concrete, objective evidence that we successfully teach poor Black children. If we instructors of teachers "purport to make a difference," we need to provide some necessary steps to breach the walls of social inequity.

KEYWORDS

- **Diversity**
- **Multiculturalism**
- **Racism**
- **Social justice**

REFERENCE

Adams, M.; Blumenfeld, W. J.; Casteñeda, C.; Hackman, H. W.; Peters, M. L.; and Zúñiga, X.; Readings for Diversity and Social Justice. 2nd Ed. New York: Routledge; **2010**.

Ayalon, A.; Why is rural education missing from multicultural education textbooks? *The Educ. Forum*. **2004,** 68(1), 24–31.

Brooks, G.; Blacks. Chicago: Third World Press; **2000**.

Delpit, L.; Other People's Children: Cultural Conflict in the Classroom. New York: New Press; **2006**.

Delpit, L.; "Multiplication is for White People": Raising Expectations for Other People's Children. New York: New Press; **2012**.

Giroux, H. A.; Theory and Resistance in Education: A Pedagogy for the Opposition. South Hadley, MA: Bergin & Garvey; **1981**.

Giroux, H. A.; Living Dangerously: Multiculturalism and the Politics of Difference. New York: P. Lang; **1991**.

Holzman, M.; and The Schott Foundation for Public Education; Yes we can: The Schott 50 State Report on Public Education and Black Males. **2010**.

Iasevoli, D.; " 'A world of white and snowy scents' ":Teaching whiteness. *Thought Action.* Fall **2009,** 57–64 p.

Killing the Sky: Oral Histories from Horizon Academy, Rikers Island, New York: Teachers College Student Press Initiative; **2005**.

Kliebard, H. M.; The Struggle for the American Curriculum, 1893–1958. 3rd Ed. New York: Routledge Falmer; **2004**.

Kozol, J.; The Shame of the Nation: The Restoration of Apartheid Schooling in America. New York: Crown; **2005**.

CHAPTER 8

CONCEPTUALIZING MALE COLLEGE STUDENTS AS MEN: A COUNSELING PERSPECTIVE WITH IMPLICATIONS FOR PROFESSIONALS IN HIGHER EDUCATION

CLAUDE M. ALDOUS and BEVERLY A. BURNELL

CONTENTS

8.1 INTRODUCTION

College men have become, in many ways, an at-risk group on college campuses. Concerns have been identified regarding recruitment and retention of male students, the greater likelihood of male students struggling to succeed academically, the greater likelihood of college men being the victims and perpetrators of violence—including suicide, higher rates of college men violating college policies, and increasing rates of college men exhibiting decreased emotional, mental, and physical well-being, including substance use and abuse (Aud, Wilkinson-Flicker, Kristapovich, Rathbun, Wang, and Zhang, 2013; Edwards and Jones, 2009; Groeschel, Wester, and Sedivy, 2010; Shen-Miller, Isacco, Davies, St. Jean, and Phan, 2013). Scholars have investigated the relationship between these aspects of college men's lives and the fact that these students are *men*, acting as they believe they are expected to as *men* in both the college context and in the broader culture of the United States. In a review of 30 years of research on men's gender role conflict, O'Neil notes that "restrictive ways of thinking about masculine norms are significantly correlated with men's psychological problems and interpersonal conflicts" (2013, p. 490). This chapter challenges higher education professionals to understand college men as men; men who have been and are continuing to learn what it means to be men within a social context that, they believe, stipulates their behaviors as college students, as residence hall members, as partners in friendships and romantic relationships, as athletes, and as members of their college and home communities.

8.2 MASCULINITY AND MALE STUDENTS

A common critique of past research about college students in the social sciences and the field of student affairs is that earlier research tended to focus upon White men, while excluding women and other nondominant groups who were far less represented in higher education at the time. The fact is that conclusions specific to gender cannot be culled from research that does not specifically examine gender. In the absence of a working conceptualization of male gender, student affairs professionals risk both alienating male students and perpetuating stereotypes about male behavior (Laker, 2003).

Feminist and social constructionist explorations of gender have produced a plethora of lenses with which to examine "masculinity" as a construct. While recent conceptualizations view gender "performatively," that is, as performance—something one *does* (Addis, Mansfield, and Syzdek, 2010), other views are essentialist, viewing gender more akin to biological sex, something one *is*, acknowledging differences between the sexes as stemming from innate biological differences (Brooks, 2001). A social constructionist conceptualization of masculinity is the focus of this chapter. This view recognizes the socially formulated and maintained nature of gender. In keeping with this perspective, it is of use to comprehend the social expectations biological males tend to face in the United States. Men in US culture are stereotypically expected to be competitive, resilient, emotionally restricted (with the exception of anger), stoic, confident, virile (sexually and otherwise), and to perform as "breadwinners," or providers. More negative expectations abound as well, as a stereotypically "real" man is antifeminine, heterosexual (often homophobic), and violent (as in "warrior" military, hunting, and sports culture). Male children are implicitly and explicitly socialized from birth into adulthood with these or similar messages about the *what* and *how* of "being a man" (Vandello and Bosson, 2012). Men in higher education settings serving as faculty, staff, or students are not exempt from these expectations.

The state of "being a man" has been described as both difficult to attain, and difficult to maintain (Vandello and Bosson, 2012). With respect to the performativity and social construction of gender, the study of multiple possible masculinities has emerged as related to, yet (at least nominally) distinct from, "gender studies." It has been noted that the word "gender" is often equated to refer to "women" (Edwards and Jones, 2009), much as how "race" matters stereotypically refers to people of color and excludes whites. Qualitative researchers have begun to examine the "lived experience" of how some college men "wear a mask" as they navigate societal, familial, and peer expectations, while privately developing their own gender identities (Edwards and Jones, 2009). This distinction between public and private gender identities is quite consistent with the concept of gender as performance. It is the "lived experience" of individuals that serves as important data when one works closely with individuals, as is often the case across higher education settings.

As is the case with all higher education professions, it is important for counseling practitioners to remain abreast of research in the field. In

Watson and Nathan's (2008) exploration of gender in the outcomes of individual and group cognitive-behavioral therapy, gender was not shown to mediate therapy outcomes. Taken alone, this might suggest to the practitioner that attending to the gender of male clients need not be a focus of therapeutic work. Given the structural limitations upon clinician time and increased severity of presenting concerns at college counseling centers (Barr, Krylowicz, Reetz, Mistler, and Rando, 2011), many clinicians might indeed reach this conclusion. The risk inherent in such a conclusion is the creation of a blind spot for therapists, and the maintenance of a pattern of service underutilization by male students. A recent survey of college counseling center directors revealed that 52.6 per cent of item respondents (N = 130) considered males an underserved population (Mistler, Reetz, Krylowicz, and Barr, 2012). More recent research has brought attention to such interventions as a diagnostic schema for assessing an individual man's masculine ideology and related constructs within the therapy setting (O'Neil, 2013), and an ecological and collaborative model for working with college men within the natural environments in which they live and study on college campuses (Shen-Miller et al., 2013). The value of attending to gender in therapeutic work with men cannot be understated, particularly given the fact that brick and mortar campuses function as communities. Male students with increased awareness of their gender socialization are better prepared to contribute positively to their campus communities.

8.3 MASCULINITY AND HEALTH

Calls for further research to inform both understanding and treatment of males with respect to gender exist in both medical and mental health fields (Meryn, 2005; Rochlen, 2005). Good, Thomson, and Brathwaite (2005) sounded the call for practitioners to exercise awareness of the growing conceptualization of multiple masculinities as a part of gender-conscious work with men.

Men are known to have higher mortality rates than women in all age groups, except those over the age of 75 years, and, as has been noted, are known to be reluctant to seek health services (Kilmartin, 2001). Given the social proscriptions against men seeking help and simultaneous prescription that men must be both strong and independent, this phenomenon is hardly surprising (Vogel, Heimerdinger-Edwards, Hammer, and Hubbard,

2011). In a pilot study by Smith, Robertson, and Houghton (2006), 80 per cent of physicians surveyed responded favorably to the idea of including a mental health component with men's annual physical assessment and indicated they would appreciate the support of counselors in this process. While the sample was small ($N = 125$), the results speak to the importance of men's mental health from the perspective of physicians. The inclusion of a mental health assessment in this annual examination represents an attempt to make efficient use of time with men who may be less likely to seek help for psychological reasons.

With respect to men's mental health, higher degrees of male role adherence are strongly associated with decreased help-seeking behaviors, a phenomenon that the mental health world shares with medicine, where it has been documented that men tend to make physician visits less frequently than do women (Brooks, 2001). A distinction from the world of physical health is the self-stigma that men may experience in seeking help for mental health issues. Self-stigma, or the shaming sense that one should not need external help, has been shown to significantly limit help-seeking, and is more prominent the more traditionally "male" a man sees himself (Vogel, Heimerdinger-Edwards, Hammer, and Hubbard, 2011; Vogel, Heimerdinger-Edwards, 2012). Therefore, significant is this issue that the National Institute of Mental Health has endorsed a "Real Men, Real Depression" campaign to raise public awareness of men's mental health issues (National Institute of Mental Health, 2003).

8.4 GENDER IN PSYCHOPATHOLOGY

Distress, as experienced *by* men, is valuable to examine. What is of additional and present interest is the experience of distress *for* men. Traditional, or hegemonic, masculine gender roles for men may conflict with modern expectations, such as having increased involvement with parenting, compared with previous generations (Ridge, Emslie, and White, 2010). While fathers may be comfortable in this role, the social expectations of traditional gender roles still exist, which prohibit many nurturing tasks of parenting as feminine. Research has demonstrated a subtle "backlash" against men who transgress the status quo of certain gender roles (Moss-Racusin, Phelan, and Rudman, 2010). Men navigate the gender world in a state of precarious balance and conflict. Being publically seen pushing a stroller and carrying a diaper bag, a typical parenting activity, may be a source of stress for men who endorse hegemonic masculinity.

Considering again the proscription against help-seeking and emotional expression, men often cope with distress using avoidance strategies as a way to preserve their manhood. This common coping strategy contributes to the delayed presentation of distress, which could contribute to professionals overlooking early or impending mental health concerns (Ridge, Emslie, and White, 2010). Men's gender role conflicts have been identified as men's experiences in three realms: devaluations, restrictions, and violations, each of which can be experienced as within the man himself, as caused by others, or expressed toward others (O'Neil, 2013). Even a gender role conflicted "modern" man may be as reluctant to seek help as a man who presents a traditional hegemonic masculinity, as he struggles to preserve his sense of masculinity against perceived infractions.

Research with male American veterans has demonstrated that conformity to male gender norms is related significantly to post-traumatic stress disorder symptoms (Morrison, 2011) and overall decreased psychological well-being (Alfred, Hammer, and Good, 2013). While the predominantly male "warrior" culture of the armed forces may contribute to military objectives being met and missions accomplished, it is of concern that male veterans may be disadvantaged in adequately coping with their experiences in combat. Male veterans risk being stigmatized for violating the "man code" of the military, and thereby risk alienation from their immediate support system if they seek help or acknowledge distress. It has been hypothesized that emotional inexpressivity and subsequent lack of adequate emotional processing of trauma may play a significant role in the genesis and maintenance of what is commonly described as post-traumatic stress disorder (Morrison, 2011). If this is even partially the case it speaks to the need for therapists working with male clients to acknowledge the role of a man's adherence to traditional gender expectations in therapy (O'Neil, 2013; Shen-Miller et al., 2013).

Restrictive gender roles have been associated with increased psychopathology for men. This is telling when one considers the resounding feminist critique of the diagnostic systems in mental health which, it is contended, diagnoses women with psychopathology disproportionately more than men. Men's physical health, too, is affected by restrictive gender roles, particularly with regard to help-seeking behaviors (Brooks, 2001). Masculinity has also been shown to affect cardiac-related health behaviors in men (Morrison, 2011). Gender roles directly affect the health and wellness of men and do so in higher education settings as well. While the majority of male students may not be fathers or veterans, adherence to rigid gender performance in adaptation to inflexible hegemonic gender roles can increase student risk in the areas of physical and mental health (Groeschel, Wester, and Sedivy, 2010 Shen-Miller et al., 2013).

8.5 CHALLENGES IN THERAPY

Men who adhere to traditional masculinities face difficulties in the realm of personal therapy, difficulties that stem from the very traits that we see associated with the development and maintenance of their presenting concerns. Feelings of shame and discomfort from needing and seeking help stand to be compounded when the rules of masculinity have disallowed emotional expression—a central component in most systems of psychotherapy. Men who may be accustomed to, and expect, a degree of "professional" distance could be uncomfortable with the degree of personal disclosure expected in the therapeutic setting. Therapists must be able to maintain flexibility to critically consider the appropriateness of their therapeutic approach, and find new ways of working with traditional males when sleeves are rolled up and the work of therapy begins (Gillon, 2008; O'Neil, 2013; Shen-Miller et al., 2013).

Although one might expect that male therapists would have an advantage working with male clients, the fact is that male therapists must be precisely aware of the effect of their own gender on the therapeutic alliance. All of the disadvantages of the normative male gender role exist in the therapy setting, including homophobia, and proscriptions against revealing "weakness" or emotions to other men also potentially exist. These socially constructed ways of performing gender are at play both within and between male clients and therapists. Male clients, presenting in a vulnerable state, may experience a great deal of difficulty accepting the help of a therapist, or articulating a presenting concern. It is of little surprise that men often seek therapy as a last resort, when doing so is a violation of traditional male roles, resulting in distress of its own accord (Ipsaro, 1986). Faculty and professional staff working in nonclinical settings in higher education can appreciate gender-related challenges male students face in the areas of help-seeking and forming relationships with the professionals endeavoring to assist them. These challenges exist in the classroom, residence halls, athletic fields, and at the administrative level.

8.6 ASSESSMENT AND TREATMENT

In an examination of evidence-based assessment with men, Cochran (2005) notes the paucity of research-based evidence regarding the psychology of men. In the absence of such evidence, Cochran encourages the use of recommendations based on clinical expertise, client values or preferences, as well as recommending research in this area for clinicians working with men. He illuminates previous work related to men's gender-role conflict cognitions, relationship patterns, men's depression, and male alexithymia (inability or difficulty in identifying emotional states) as areas for assessment work with men. Statistically, men are most likely to experience diagnosable mental disorders related to alcohol and substance abuse, followed by antisocial personality disorder, then depression. Considering the dearth of evidence-based assessment tools, it is important that clinicians develop an informed manner of assessment with male clients which informs both case conceptualization and treatment planning. O'Neil's diagnostic schema (2013) is a recent contribution with promise for identifying correlations between four patterns of male gender role conflict and men's mental health concerns. O'Neil further asserts that "respecting and affirming the client's own views of his problems is critical in developing the therapeutic alliance and setting the stage for the deepening and the internal emotional work that is often necessary" (2013, p. 497). The theory-into-practice nature of this evolving field of study requires both researchers and clinicians to continue to critically examine the extant research and remain abreast of the development of new strategies to better serve men in therapy (Cochran, 2005).

Kilmartin (2005) calls for therapists to "teach about gender" (p. 97) when working with male clients, and draws attention to the nature of the current diagnostic system to prioritize women's tendency to "act-in" regarding depression. He makes a case for conceptualizing a masculine form of depression, and provides useful ideas for treatment centered upon the use of skill building and an expansion of positive masculine traits, such as facing challenges, courage, assertiveness, and independence. The reframing of help-seeking and treatment as existing within a masculine context has the potential to increase men's adherence to treatment and to increase male gender awareness. These gender-aware interventions all serve the psycho-educational purpose of teaching male clients new, more adaptive ways of "doing" masculinity (Kilmartin, 2005). The reframing of help-seeking and expansion of positive masculine traits has utility in various higher education settings as well. Gender awareness can enrich interventions and interactions with students and, when appropriate, can serve as standalone interventions with potentially broad applicability.

Traditional masculinity disadvantages men in the therapeutic setting, which can create discomfort and attrition from treatment. Emotional restrictiveness, shame at needing help, and reticence about the process of therapy are some of the impediments men face at the onset of therapy. Motivational interviewing conceptualizations and techniques hold promise for reluctant male clients. There is also a need to tailor interventions specifically to address areas of risk (regarding alcohol and substance abuse, medical avoidance, etc.), all while utilizing conceptualizations that acknowledge the role of masculinity in the client's life. Group interventions hold a unique promise for depotentiating men's competitive socialization, ideally creating a climate of interdependence that may be unique in the lives of many men, who may be unaccustomed to relying on other men for overt emotional support (Good, Thomson, and Brathwaite, 2005).

8.7 WHERE THE WHEELS MEET THE ROAD

To work effectively with male clients, it is essential for therapists to develop a clinical repertoire that acknowledges the place of gender in the etiology of men's concerns, as well as in the treatment of their concerns. Such conceptualizations serve the dual purpose of identifying potent areas for change, as well as increasing client awareness of self. Gender informed therapists can tailor specific interventions utilizing the strengths and preferences of male clients, while encouraging them to seek new and more effective ways of "doing" masculinity. Brooks (2001), for example, states that men's "abilities to function as nurturers and caregivers must be given far greater emphasis" (p. 33).

Therapeutic and higher education practice with male students has much to gain from a gendered examination of men, their concerns in seeking help, and appropriate interventions which honor men as unique individuals, with unique strengths and resiliencies, as well as the challenges of "performing masculinity."

In the preparation of this manuscript, the male author (Claude) experienced some interesting personal revelations. My conceptualization of my own masculinity has changed, or rather, my understanding of how I typically "do" masculinity has evolved. The accumulation of evidence, upon reflection, supports that I am significantly more competitive and confrontational than I had previously thought—both prominent and traditionally masculine traits. In considering myself somewhat progressive, particularly for having origins in a conservative, rural area, I had often felt disinterested in competition with others for its own sake, and strongly preferred to follow my own interests. I realize now that I had, in many ways, sought to distinguish myself from expressions of masculinity that (at the time) I found boorish and boring. However, despite these efforts, my style of presentation, in dress, speech, and interests were, and are, relatively easily framed in the context of competition and confrontation, both hegemonic masculine traits. The distinction I sought between myself and other men was about how I "did" masculinity, not about transcending or up-ending what felt like inappropriate expressions or distasteful masculine behaviors. Asserting my independence from others, interestingly, fits rather neatly into the context of hegemonic masculinity, perhaps the very concept I sought distance from. In coming to view the distinction in terms of performance, the effect is that I effectively close some interpersonal distance between myself and other men, having navigated similar expectations using similar masculine traits.

As a new professional counselor working in a college counseling center, having completed training experiences in college counseling, as well as a graduate assistantship in a residence life department, I am afforded a degree of perspective on this topic. My hope is that as a therapist I will be more effective with male clients as I have become better able to conceptualize gender in my own life. I find that when working with male clients, I can often frame their presenting concerns, previous coping techniques, and worldviews in the context of masculinity in a manner that is straightforward and matter-of-fact, but also validating, normalizing, and encouraging. Respect for the male client is paramount and works against what Laker (2003) would call the " 'Bad Dog' approach" (p. 2), which endeavors solely to correct behavior, often at the expense of deeper, developmental interventions. Approaching college men with sensitivity and respect rather than a corrective, and potentially scolding orientation can set the stage for meaningful change, and growth. With female students, it can be effective to reframe aspects of the men in their lives through the lens of gender as well, which can address stereotypes about male behavior and aid their conceptualization of relational concerns with men in the variety of settings in which these women interact with males.

Executed with sensitivity and respect, these gender aware conceptualizations have the potential to be effective for both male and female professionals serving in myriad roles in higher education. If we are better able to address men's gender within higher education, we will be serving not only our male clients, but may also contribute to positive social change.

KEYWORDS

- **College men**
- **Post-traumatic stress disorder**
- **Psychopathology**

REFERENCES

Addis, M. E.; Mansfield, A. K.; and Syzdak, M. R.; Is "masculinity" a problem?: framing the effects of gendered social learning in men. *Psychol. Men Mascul.* **2010,** *11(2),* 77–90. doi: 10.1037/a0018602

Alfred, G. C.; Hammer, J. H.; and Good, G. E.; Male student veterans: Hardiness, psychological well-being, and masculine norms. *Psychol. Men Mascul.* Advance Online Publication; **2013,** ***15*** **(1), 95–99**. doi: 10.1037/a0031450

Aud, S.; Wilkinson-Flicker, S.; Kristapovich, P.; Rathbun, A.; Wang, X.; and Zhang, J.; The Condition of Education 2013 (NCES 2013-037). US Department of Education. Washington, DC: National Center for Education Statistics; Retrieved December 9, 2013 from **2013**.

Barr, V.; Krylowicz, B.; Reetz, D.; Mistler, B. J.; and Rando, R.; The Association for University and College Counseling Center Directors Annual Survey. Retrieved December 13, 2013 from http://www.aucccd.org/support/aucccd_directors_survey_monograph_2011.pdf, **2011**.

Brooks, G. R.; Masculinity and men's mental health. *J. Am. Coll. Health.* **2001,** *49(6),* 285–297. doi: 10.1080/07448480109596315

Cochran, S. V.; Evidence-based assessment with men. *J. Clin. Psychol.* **2005,** *61(6),* 649–660. doi: 10.1002/jclp.20100

Edwards, K. E.; and Jones, S. R.; "Putting my man face on": a grounded theory of college men's gender identity development. *J. Coll. Stud. Dev.* **2009,** *50(2),* 210–228. doi: 10.1353/csd.0.0063

Gillon, E.; Men, masculinity, and person-centered therapy. *Person-Cent. Exp. Psychoth.* **2008,** *7(2),* 120–134.

Good, G. E.; Thomson, D. A.; and Brathwaite, A. D.; Men and therapy: critical concepts, theoretical frameworks, and research recommendations. *J. Clin. Psychol.* **2005,** *61(6),* 699–711. doi: 10.1002/jclp.20104

Groeschel, B. L.; Wester, S. R.; and Sedivy, S. K.; Gender role conflict, alcohol, and help seeking among college men. *Psychol Men Mascul.* **2010,** *11(2)* 123–139. doi: 10:1037/a0018365

Ipsaro, A. J.; Male client–male therapist: issues in a therapeutic alliance. *Psychother: Theor. Res. Pract. Train.* **1986,** *23(2),* 257. doi: 10.37/h0085607

Kilmartin, C.; Surviving and thriving: men and physical health. In: Cohen, T. F.; Men and Masculinity. Belmont, CA: Wadsworth; **2001**, 352–360 p.

Kilmartin, C.; Depression in men: communication, diagnosis and therapy. *The J. Men's Health Gender.* **2005,** *2(1),* 95–99. doi: 0.1016/j.jmhg.2004.10.010

Laker, J.; Bad dogs: rethinking our engagement of male students. *ACPA's Stand. Committee Men Newslett.* March **2003**, 1-2.

Meryn, S.; Qualitative research, communication in sexual and men's health, and who is the man's doctor? *J. Men's Health Gender.* **2005,** *2(1),* 3–6.

Mistler, B. J.; Reetz, D. R.; Krylowicz, B.; and Barr, V.; The Association for University and College Counseling Center Directors Annual Survey. Retrieved December 12, 2013 from http://www.aucccd.org/support/Monograph_2012_AUCCCD%20Public.pdf, **2012**.

Morrison, J. A.; Masculinity moderates the relationship between symptoms of PTSD and cardiac-related health behaviors in male veterans. *Psychol. Men Mascul.* **2011,** *13(2),* 1–8. doi: 10.1037/a0024186

Moss-Racusin, C. A.; Phelan, J. E.; and Rudman, L. A.; When men break the gender rules: Status incongruity and backlash against modest men. *Psychol. Men Mascul.* **2010,** *11(2),* 140. doi: 10.1037/a0018093

National Institute of Mental Health. Real Men, Real Depression. Washington DC: Author; **2003**.

O'Neil, J. M.; Gender role conflict research 30 years later: an evidence-based diagnostic schema to assess boys and men in counseling. *J. Couns. Dev.* **2013,** *91,* 490–498. doi: 10.1002/j.1556-6676.2013.00122.x

Ridge, D.; Emslie, C.; and White, A.; Understanding how men experience, express, and cope with mental distress: Where next? *Soc. Health Illn.* **2011,** *33(1),* 145–159. doi: 10.1111/j.1467-9566.2010.01266.x

Rochlen, A. B.; Men in (and out of) therapy: Central concepts, emerging directions, and remaining challenges. *J. Clin. Psychol.* **2005,** *61(6),* 627–631. doi: 10.1002/jclp.20098

Shen-Miller, D. S.; Isacco, A.; Davies, J. A.; St. Jean, M.; and Phan, J. L.; The men's center approach: Ecological interventions for college men's health. *J. Coun. Dev.* **2013,** *91,* 499–507. doi: 10.1002/j.1556-6676.2013.00123.x

Smith, J. M.; Robertson, S.; and Houghton, B.; Physicians' views on men's mental health and the role of counselling services: a pilot study. *Coun. Psychother. Res.* **2006,** *6,* 258–263. doi: 10.1080/14733140601029241

Vandello, J. A.; and Bosson, J. K.; Hard won and easily lost: a review and synthesis of theory and research on precarious manhood. *Psychol. Men Mascul.* **2012,** *14(2),* 101–113. doi: 10.1037/a0029826

Vogel, D. L.; Heimerdinger-Edwards, S. R.; Hammer, J. H.; and Hubbard, A.; "Boys don't cry": examination of the links between endorsement of masculine norms, self-stigma, and help-seeking attitudes for men from diverse backgrounds. *J. Coun. Psychol.* **2011,** *58(3),* 368. doi: 10.1037/a0023688

Watson, H. J.; and Nathan, P. R.; Role of gender in depressive disorder outcome for individual and group cognitive-behavioral treatment. *J. Clin. Psychol.* **2008,** *64(12),* 1323–1337. doi: 10.1002/jclp.20524

CHAPTER 9

CULTURAL COMPETENCY DEVELOPMENT FOR NEW STUDENT AFFAIRS PROFESSIONALS

JELANE A. KENNEDY, WENDY NEIFELD WHEELER, and STEPHANIE BENNETT

CONTENTS

9.1 INTRODUCTION

The demographic profile of college campuses is more diverse today, in the twenty-first century, than at any other time in higher education history (US Bureau of Census, 2010). In the 2011 Minorities in Higher Education, Kim (2011) reported "the past two decades witnessed a steady rise in college enrollment rates among young people in every racial/ethnic background" (p. 2). As the student population continues to evolve and become more varied, student affairs professionals have worked to keep pace with the knowledge and level of competency required to provide expanded and new multicultural approaches to traditional services. Promoting and cultivating an environment that fosters belonging and inclusion is essential for all student services practitioners. Pope and Mueller (2005) suggest that "the incorporation of multicultural knowledge as well as exploration of multicultural awareness and skills within student personnel preparation programs is vital to the development of student affairs practitioners who are competent to work ethically and effectively with all students" (p. 680). Shuford (2011) echoes, "higher education needs professionals that possess the knowledge, skills, and awareness needed to foster cross-cultural understanding and engagement among diverse student populations on campus" (p. 247).

During the past five decades, the field of student services has conveyed a commitment to the ideals of social justice, cultural competency, and advancing multicultural education (Burkard, Cole, Ott, and Stoflet, 2005; Pope, Mueller, and Reynolds, 2009). Informed by practice-oriented inquiry and scholarship, student affairs practitioners are often called upon to enhance and institutionalize campus diversity (Howard-Hamilton, 2000; Mueller and Pope, 2001; Pope et al., 2009; Waple, 2006). The significance of multicultural understanding, training, and skills for student affairs professionals has been formally acknowledged by the American College Personnel Association (ACPA: College Student Educators International), the National Association of Student Personnel Administrators (NASPA: Student Affairs Administrators in Higher Education), and the Council for the Advancement of Standards in Higher Education (Pope and Reynolds, 1997; Professional Competency Areas for Student Affairs Practitioners, 2010; Talbot, 1996b). In 2010, ACPA and NASPA published a joint document of professional competency areas for student affairs that clearly outlines the expressed need for entry level practitioners to have developed "knowledge, skills, and attitudes" related to cultural competency as part of the section titled Equity, Diversity, and Inclusion.

In examining the Council for the Advancement of Standards in Higher Education (2012) for master's programs in Student Affairs Administration and the Council for the Advancement of Higher Education Programs (2009), it is noted that program designers are encouraged to expose students to cultural identity theory and to examine characteristics about a variety of student populations including those marginalized by society as a whole. Mclellan and Larimore (2009) emphasize that "as student affairs practitioners, we must rededicate ourselves to quantitative change while simultaneously committing ourselves to qualitative change in the lived experiences of our students if we are to advance the interest of assuring the success of our students and our institutions" (p. 237). This is a call to graduate students to understand and explore what it personally means to be a racial/cultural being (i.e., to unpack one's biases and prejudice, to have an understanding of the impact self-awareness can have on marginalized students, and to be sensitized to others' world views) and move beyond the superficial and common "intellectual-only" process of understanding the student experience (Flowers and Howard-Hamilton, 2002, McEwen and Roper, 1994; Pope et. al., 2009). Reynolds and Rivera (2012) suggest that the "model of multicultural awareness, knowledge and skills be expanded to include other dimensions affecting one's ability to work with diverse populations such as awareness of one's own assumptions and biases" (p. 167).

Despite what seems to be a genuine concerted effort to prepare entry-level student affairs practitioners to work effectively with diverse populations, Pope and Reynolds (1997) recognized almost 15 years ago that there is a lack of required systematic training and exposure in the development of cultural competencies as part of student affairs graduate programs. In 2003, Flowers used the Directory of Graduate Programs: Preparing Student Affairs Professionals (authored by Coomes and Talbot, 2000) to examine the diversity requirements in student affairs programs. From the directory, programs that met the criteria of having one full-time faculty member; four content courses (minimally) about student services; a required field experience component; and a 2-year program design were selected to participate in the study (Flowers, 2003). Fifty-three program administrators completed the survey and, of those, 74 per cent indicated that there was a required diversity course as part of the core curriculum.

To determine if any change had occurred since the Flowers' (2003) study, the current authors investigated the public information available about master's level graduate programs in student services and higher education administration in Fall 2012. The examination included accessing data from the Association of College Personnel Administrators (ACPA) online directory of graduate programs (2012), the National Association of Student Personnel Administrators (NASPA) online directory of graduate programs (2012), the Association for the Study of Higher Education (ASHE) online directory of graduate programs (2012) and the Council for Accreditation of Counseling & Related Educational Programs (CACREP) online directory (2012) of graduate programs in student affairs and higher education administration. From these records, it was determined that there are currently 223 master's level graduate programs in student services and higher education administration programs in the United States. Of these programs, only 43 per cent included a diversity course requirement as part of the core curriculum. Despite what continues to be overt direction from our national professional organizations to have robust knowledge, attitude, and skills as it relates to diversity and cultural competency, more than half of our graduate preparation programs fail to require or integrate coursework relative to multicultural competence into the curriculum.

9.2 PURPOSE OF PRESENT STUDY

The purpose of this exploratory research, which was part of a more comprehensive project, was to examine the influence of a required social and cultural foundations course on the development of cultural competency skills of graduate students pursuing a master's degree in student services at a predominately White institution. Using a pre-test and post-test multicultural competency assessment instrument, the investigators measured knowledge, awareness, and skill development.

The hypotheses were the following:

H1: There is a significant change between individual students' level of cultural competence (awareness, knowledge, and skills) from the beginning to the end of the semester of a required Social and Cultural Foundations course.

H2: There is a significant change between individual students' level of cultural competence (awareness, knowledge, and skills) from the beginning to the end of the semester of a required Social and Cultural Foundations regardless of which of the five class sections the student participated in.

9.3 METHOD

9.3.1 SAMPLE

Participants represented a subset of a total of 96 graduate counseling and student services students enrolled in five sections of a Social and Cultural Foundations course that occurred over three semesters. One student dropped the course, two students did not fully complete the assessments, and three students chose not to participate in the study. Of the 90 remaining participants, 21 were enrolled in a noncohort model graduate student services program (16 women and 5 men). The analysis of data includes only those individuals matriculated in the student services program. The principle investigator taught all five sections of the Social and Cultural Foundations class utilizing the same syllabus, textbooks, and course activities. All students in the student services program were enrolled at a small private liberal arts college in the Northeast and were pursuing a Master's degree in education. A convenience sample was used; students self-selected into one of the five sections that met the student's scheduling needs over a period of three semesters. Students were not told of the study until after enrollment in the classes.

9.3.2 THE SOCIAL AND CULTURAL FOUNDATIONS COURSE

The five classes met for 2.5 h once a week over 15 weeks for a total of 37.5 h of instructional time. All five classes followed the identical format with homework assignments, guest speakers, DVDs, readings from textbooks and classroom activities [i.e., small and large group discussion, lecture and a participation in the simulation BARNGA, a card game that allowed students to explore cultural differences (Neville, Heppner, Louie, Thompson, Brooks, and Baker, 1996)]. The students were asked to select a focus group for the semester. The focus group was to be different from the student in some way (i.e., race, religion, sexual orientation, etc.) and was vetted with the instructor through conversation in each class. As a writing-intensive class, the students completed (1) weekly reflective journals and awareness activities; (2) three guided interaction papers related to the focus group choice—each interaction required higher levels of contact with the focus group; (3) a book report that the students shared with classmates related to the focus group; (4) a prescribed culture paper that asked students to compare and contrast personal views on one's own culture with cultural and religious specific readings in two of the required textbooks; and (5) a final guided reflection paper that assessed overall growth as a more culturally competent professional. The first three class meetings emphasized building rapport, creating a safe environment, and setting ground rules for the class (Ramsey, 1999). This was done with icebreaker activities that explored cultural sameness and differences. In two of the three classes students generated ground rules for creating a safe space that would be used throughout the semester.

This course was designed to promote an awareness of one's own cultural biases, to expand the knowledge base about different cultural groups, to enhance practical and professional skills to make culturally appropriate interventions, and to encourage basic to complex immersion opportunities. The instructor comes from a multiculturalist and feminist teaching viewpoint. This foundation has guided the course curriculum content that has expanded beyond focus on the traditional four racial minority groups and discussions of White privilege to include topics related to sexual orientation, gender, religion, class/socioeconomic status, national origin and disability status (Gayles and Kelly, 2007; Priester et al., 2008; Smith-Adcock, Ropers-Huilman, and Choate, 2004).

The instructor of the class was a seasoned professor who had designed and taught the class for 14 years. The instructor's research interests have included the impact of classroom diversity on the learning environment and individual student's progress in multicultural education. In respect for full disclosure, the instructor identifies as European American, female, and as a lesbian.

9.3.3 DEMOGRAPHICS

The students in all five classes attend a predominantly White institution. Students enrolled in the student services program ($n = 21$) were 76 per cent female (n = 16) and 24 per cent male (n = 5). The ages of the participants ranged from 20 to 39 years old; 20–24 (52%, n = 11), 25–29 (29%, n = 6), 30–34 (14%, n = 3), and 35–39 (5%, n = 1). The 21 participants self-identified as 76 per cent (n = 16) European American and 24 per cent (n = 5) as students of color. In examining religion, 90 per cent (n = 19) identified as Christian, and 10 per cent (n = 2) as non-Christian. Finally, in regards to sexual orientation, 81 per cent (n = 17) identified as heterosexual and 19 per cent (n = 4) a sexual minority.

It is important to note that the number of participants that self-identified from each racial/ethnic, religious, and sexual orientation category was too small to conduct useful analysis so these data were collapsed into dichotomous variables. The demographics were divided into groupings of those who identified with characteristics of the dominant culture (i.e., European American, Christian, and heterosexual) and those who did identify as belonging to a traditionally marginalized group (i.e., students of color, non-Christian [Jewish, Muslim, nonbelievers, no religious upbringing and atheist] and sexual minority [bisexual, gay/lesbian, and transgendered]) Age and gender were not collapsed.

9.4 INSTRUMENT

9.4.1 MULTICULTURAL AWARENESS KNOWLEDGE SKILLS SURVEY-COUNSELOR FORM (MAKSS-C)

D'Andrea, Daniels, and Heck (1991) created the MAKSS-C to assess students' multicultural competence. The MAKSS-C was chosen over the MAKSS-CE-R after consultation with the first author of the instrument, Dr. Michael D'Andrea. The MAKSS-C and MAKSS-CE-R are measures that were developed to assess the multicultural counseling competencies based on the model of cross-cultural counseling by Sue and his colleagues (Sue, Bernier, Durran, Feinberg, Pedersen, and Smith, 1982). According to Dr. D'Andrea (personal communication, February 10, 2009) the MAKSS-C is longer than MAKSS-CE-R and was described as better suited for use with students as a tool to measure competencies and develop students' own self-awareness. The MAKSS-CE-R was described as better utilized as a research tool—not for self-assessment. In other words, by using the MAKSS-C, students would continue to gain the benefit of thinking about their own sense of self-awareness. The MAKSS-C has 60 items that give a total score and three subscale scores. The three subscales are awareness (20 items), knowledge (20 items), and skills (20 items). Students respond by using three different 4-point Likert scales. Example of items from the MAKSS-C include awareness: "Ambiguity and stress often result from multicultural situations because people are not sure what to expect from each other" answered with a scale from Not Sure to Strongly Agree; knowledge: "The difficulty with the concept of 'integration' is its implicit bias in favor of the dominant culture" answered with a scale from Not Sure to Strongly Agree; and skills: "How well would you rate your ability to analyze a culture into its component parts?" answered with a scale Not Sure to Very Good. Students are considered to have higher competencies if their scores are higher in each subscale.

The MAKSS-C was selected because of its extensive history of use in the assessment of multicultural competence, particularly with individuals studying or practicing counseling. As this is a vital skill for emerging student services professionals, it was deemed an appropriate choice for this study. The instrument has been found to be a reliable and valid measure of multicultural competency with Cronbach's alpha reliability coefficients of awareness: 75, knowledge: 90, and skills: 96 (D'Andrea et al., 1991).

9.5 PROCEDURE

Once approval from the college's institutional review board on human studies was obtained, the principle investigator introduced the study to the five classes. Students were informed that participation was voluntary, that the students were free to refuse participation, and that it would not effect students grades negatively or positively. The students were also informed that they could withdraw from the study at any time without adverse affects on final grades in the class. Students' demonstrated their agreement by signing informed consent forms that included age and program of study. Students who agreed to participate allowed the investigator to collect their pre-test and post-test MAKSS-C and to utilize their written work (culture papers) to gain demographic information about the students. Utilizing the culture paper to collect demographic information would allow a student to express freely how he/she identifies himself or herself. Students also knew that if the information given in their culture papers was not clear, the principal investigator would meet with the student to address any confusion as to how the student identified.

9.6 RESULTS

Study data were analyzed using descriptive statistics, a paired *t*-test and an ANOVA. To examine the effect of a precisely designed culturally competency curriculum on the development of multicultural competencies, differences in pre-and post-test scores on each of the subscales (awareness, knowledge, and skills) of the MAKSS-C were compared. For all means, higher scores indicate self-reporting of higher levels multicultural competency. The range of possible points on each subscale is 20–80 points.

All 21 students matriculated in the student services program, who were enrolled in the Social and Cultural Foundation courses self-reported improvement in their scores on each of the MAKSS-C subscales from pre- to post-test. Student scores rose on the subscale awareness from 47.24 (standard deviation (SD) 6.61) to 57.19 (SD 4.95). These scores indicate that for the subscale awareness, the means increased from pre- to post-test and the variance decreased (demonstrating a smaller range of variability in post-test scores). Scores in the subscales knowledge and skills also increased from 40.43 (SD 7.92) to 56.38 (SD 8.13) and from 44.38 (SD 7.69) to 57.71 (SD 8.14), respectively.

A paired *t* -test analysis was conducted to investigate whether the amount of change in students' scores between pre- and post-test indicated a significant difference on any of the subscale measures. Students' scores demonstrated a significant difference on all three subscales from the beginning of the 15-week course to the end of the course: Awareness (t = 6.670; $p < 0.05$); knowledge (t = 7.479 $p < 0.05$); Skills (t = 5.223; $p < 0.05$). This provides statistical evidence that the intervention of the class material, rather than chance, influenced the increase in means from pre-test to post-test. The mean change on the subscale awareness was 9.952 (SD 6.837); knowledge, 15.952 (SD 9.775); and skills, 13.333 (SD 11.698). The greatest mean change was on the subscale knowledge.

Since the students in the existing sample were members of five distinct sections of the Social and Cultural Foundations course, who took the course during one of three academic semesters, their answers on measures of the dependent variables might have varied at pre-test. To investigate this possibility, an ANOVA was conducted. The rationale was to determine whether or not the mean pre-test scores on each of the subscales differed depending on which of the five classes or which semester the students were enrolled in. In examining the analysis between and within each class at pretest, no significant differences were found for any of the three subscales: awareness (F = 0.718); knowledge (F = 0.475); and skills (F = 0.557). The lack of statistical significance in this analysis indicates that all 21 students, regardless of which semester they completed the course or where they were in the progress of their academic career, entered the Social and Cultural Foundations course with comparable levels of cultural competency as measured by the MAKSS-C.

In summary, since all students demonstrated similar levels of cultural competency at the start of the course and then experienced statistically significant increases in the mean scores on the each of the MAKSS-C subscales, it appears that engaging in this course was the major catalyst for change.

9.7 LIMITS

Limitations should be considered with respect to the findings of this study. It is important to acknowledge that becoming a culturally competent student services practitioner is an ongoing process. A one-semester multicultural training course may serve as a partial component of long-term multicultural development, but enduring improvement cannot be confirmed as part of a 15-week course. And, because only self-reported data were collected, it is possible that students may have inflated their responses due to social desirability. As noted, self-reported responses may also reflect anticipated rather than actual behaviors and attitudes. The use of a non-randomized convenience sample with representation of participants from only one institution impacts the ability to generalize the results. A small sample size meant the variables had to be reported as dichotomous (dominate culture vs. non-dominant culture) and not as individual variables as noted earlier in the paper. The instrument selected was designed primarily to assess graduate students in a counseling program and specifically for students matriculated in a student services program. Finally, as noted earlier, the principal investigator was the instructor of all five classes; while this may have assisted with consistency between the classes, it may have also impacted how students self-reported on the assessment instrument. In addition, the principal investigator was aware of potential basis and utilized journaling, colleague consultation, and vigilant self-awareness to manage potential bias.

9.8 DISCUSSION

As part of a larger comprehensive investigation, this study sought to examine if individuals in a student services graduate program enhanced their cultural competency knowledge, awareness, and skills, thereby following active participation and engagement in a social and cultural foundations course. Results indicated that participation in the course appeared to boost cultural competency regardless of which section of the class the students took. This was also true irrelevant of at what point in the students' program they matriculated into the course. This confirmed the researchers' prediction that there would be a significant change between an individual student's level of cultural competence (awareness, knowledge, and skill) from the start of the course to the completion at the close of the semester. Statistical analysis also confirmed that there was no significant difference in MAKSS pre-test scores across participants in the five scheduled classes.

9.9 IMPLICATIONS AND CONCLUSION

Student services and higher education practitioners are often called upon to address issues related to cultural diversity and to serve as allies for social justice (Bukard et. al., 2005; Mueller and Pope, 2001; Waple, 2006). In addition, they are frequently expected to research, identify, and develop best practices related to diversity initiatives (Castellanos, Gloria, Mayorga, and Salas, 2007; Howard-Hamilton, 2000). Despite evidence that academic communities continue to become globalized and institutions have revised mission statements to explicitly voice support for multiculturalism, there continues to be a struggle to develop genuine environments that both value diversity and encourage meaningful dialogue between diverse groups.

It is surprising that every student services and higher education administration program does not have a cultural competency course as part of a core requirement (Flowers and Howard-Hamilton, 2002; McEwen and Roper, 1994; Pope and Mueller, 2005; Pope and Reynolds, 1997).

Emerging professionals in student services and higher education administration programs are not exposed to systematic opportunities that allow engagement in understanding who they are as cultural beings (Pope and Reynolds, 1997; Reason and Broido, 2005). Reason and Broido argue that "self-understanding ... provides the foundation on which sustainable ally development and actions are built" (2005, p. 82). At minimum, programs should require at least one course where students are asked to examine their self-awareness, knowledge, and skill as related to cultural diversity. Additionally, programs should integrate diversity issues throughout course work (Flowers and Howard-Hamilton, 2002; Herdlein, Kline, Boquard, and Haddad, 2010).

One unique factor that may influence curriculum development and design may be lack of personal comfort in addressing issues of cultural competency by faculty in student services and higher education administration programs (Pope and Mueller, 2005; Reynolds, 2009; Talbot, 1996b; Talbot and Kocarke, 1997). It is reasonable to expect that both full-time and adjunct faculty members will continue to develop and work on their own cultural competencies as educators and role models to emerging professionals. Their experiences are valuable to their students (Cooper, Howard-Hamilton, and Cuyjet, 2011; King and Howard-Hamilton, 2003). This requires ongoing honest self-reflection and a commitment to one's own personal growth.

Graduate students completing our programs need to be prepared to fully engage in their work. They will be called to develop and enforce policies with an eye to cultural sensitivity; and be asked to use culturally appropriate problem-solving skills while recognizing the impact of oppression of various groups (Cuyjet, Longwell-Grice, and Molina, 2009; Kelly and Gayles, 2010). As on-campus leaders, they may be asked to be involved in assessment activities that include an evaluation of campus climate (Pope and Mueller, 2011; Reason and Broido, 2005). It is not unusual for young professionals to be asked to lead interdialogue groups addressing issues of cultural diversity and to be involved in safe zone or similar training programs (Howard-Hamilton, 2000; Kelly and Gayles, 2010; Reason and Brodio, 2005; Reason and Davis, 2005).

Lack of adequate preparation of our young professionals is a disservice to the individual and to the students they will serve. This includes the absence of self-reflection, research, and communication skills—all which can deepen their own self-awareness and cultural sensitivity (Castellanos et al., 2007; Pope and Reynolds, 1997; Reason and Broido, 2005). It is vital for faculty to prepare student services and higher education emerging professionals to be equipped to engage in diversity work on multiple levels (Pope, Reynolds, and Mueller, 2004). The success of the field of student services and higher education administration requires us to reflect on our graduate degree programs and to examine how we can best engage our students in this fast-moving and ever-changing field. Sustainability for our young professionals requires that they develop skills in life-long learning and the ability for self-care related to cultural competency. The road to being culturally competent is a life-long journey (Howard-Hamilton, 2000).

Longitudinal research on the impact of cultural competency training in graduate preparation programs is needed. Additional ideas to be explored should acknowledge that today's graduate student is technologically savvy and may benefit from different strategies of cultural competency development. Questions might include the following: What works? What do they need more of? How can we help young professionals sustain their growth? What support models would work best? How does ongoing professional development have an impact? It may be helpful for future researchers to consider available research tools such as multicultural competence in student affairs—preliminary 2 scale (MCSA-P2) and multicultural competence characteristics of student affairs professionals inventory (MCCSA-PI) in forthcoming studies. Equally, it may be important for supervisors to consider developing ongoing diversity training goals as part of a division and department commitment to continued cultural competency growth (Reynolds, 2009). Ultimately, this work can have a valuable impact on the college, community, and our global neighbors, thereby helping our new professionals to be prepared for this dynamic workplace.

KEYWORDS

- **Cultural competency**
- **Demographics**
- **Student affairs professionals**

REFERENCES

ACPA College Student Educators International. Directory of Graduate Programs Preparing Student Affairs Professionals. Retrieved from http://www2.myacpa.org/profprep-program-directory/profprep-directory-home2, **2012.**

ASHE Association for the Study of Higher Education. ASHE Higher Education Program Directory. Retrieved from http://www.ashe.ws/page.asp?page=587, **2012**.

Burkard, A. W.; Cole, D. C.; Ott, M.; and Stoflet, T.; Entry-level competencies of new student affairs professionals: a delphi study. *J. Stud. Aff. Res. Pract.* **2005,** *42(3),* 545–571. doi:

CACREP Council for Accreditation of Counseling and Related Educational Programs. CACREP's Online Interactive Directory. Retrieved from http://www.cacrep.org/directory/directory.cfm, **2012**.

Castellanos, J.; Gloria, A. M.; Mayorga, M.; and Salas, C.; Student affairs professionals' self-report of multicultural competence: understanding awareness, knowledge, and skills. *NASPA J.* **2007,** *44(4),* 643–663. Retrieved from http://journals.naspa.org/

Coomes, M. D., & Talbot, D. M. (2000) *Directory of graduate programs: Preparing student affairs professionals, 1999.* Washington, D.C.: American College Personnel Association.

Cooper, D. L.; Howard-Hamilton, M. F.; and Cuyjet, M. J.; Achieving cultural competence as a practitioner, student, or faculty member theory to practice. In: Multiculturalism on Campus: Theory, Models and Practices for Understanding Diversity and Creating Inclusion. Cuyjet, Michael J.; Howard-Hamilton, Mary F.; and Cooper, Diane L.; Eds. Sterling, Virginia: Stylus Publishing, LLC; **2011,** 401–420 p.

Council for the Advancement of Higher Education Programs. A Commitment to Quality: Guidelines for Higher Education Administration and Leadership Preparation Programs at the Masters Degree Level. Retrieved from http://www.ashe.ws/images/CAHEPLeadershipProgramGuidelines.pdf, **2009**.

Council for the Advancement of Standards in Higher Education. The book of Professional Standards for Higher Education (CAS). Washington, DC: **2012**.

Cuyjet, M. J.; Longwell-Grice, R.; and Molina, E.; Perceptions of new student affairs professionals and their supervisors regarding the application of competencies learned in preparation programs. *J. Coll. Stud. Dev.* **2009,** *50(1),* 104–119. Retrieved from http://www.jcsdonline.org/accestojcsd.html

D'Andrea, M.; Daniels, J.; and Heck, R.; Evaluating the impact of multicultural counseling training. *J. Coun. Dev.* **1991,** *70,1* 143–150. Retrieved from http://www.counseling.org/publications/journals.aspx; DOI: 10.1002/j.1556-6676.1991.tb01576.x

Flowers, L.; National study of diversity requirements in student affairs graduate programs. *NASPA J.* **2003,** *40(4),* 72–82. Retrieved from http://journals.naspa.org

Flowers, L. A.; and Howard-Hamilton, M. F.; A qualitative study of graduate students' perceptions of diversity issues in student affairs preparation programs. *J. Coll. Stud. Dev.* **2002,** *43(1),* 119–123. Retrieved from http://www.jcsdonline.org/accestojcsd.html

Gayles, J. G.; and Kelly, B. T.; Experiences with diversity in the curriculum: implications for graduate programs and student affairs practice. *NASPA J.* **2007,** *44(1),* 193–208. Retrieved from http://journals.naspa.org/

Herdlein, R.; Kline, K.; Boquard, B.; and Haddad, V.; A survey of faculty perceptions of learning outcomes in master's level preparation programs in higher education and student affairs. *Coll. Stud. Aff. J.* **2010,** *29(1),* 33–45. *Retrieved from http://www.infoagepub.com/college-student-affairs-journal.html*

Howard-Hamilton, M.; Programming for multicultural competencies. *New Direct. Stud. Serv.* **2000,** *90,* 67–78. Doi: 10.1002/ss.9006 [NOTE: THIS IS A MONOGRAPH SO THERE IS NO ISSUE NUMBER]

Kelly, B. T.; and Gayles, J. G.; Resistance to racial/ethnic dialog in graduate preparation programs: implications for developing cultural competence. *Coll. Stud. Aff. J.* **2010,** *29(1),* 75–85. *Retrieved from http://www.infoagepub.com/college-student-affairs-journal.html*

Kim, Y. M.; Minorities in Higher Education: Twenty-Fourth Status Report (2011 Supplement). Washington, DC: American Council on Education; **2011**.

King, P. M.; and Howard-Hamilton, M.; An assessment of multicultural competence. *NASPA J.* **2003,** *40(2),* 119–133. Retrieved from http://journals.naspa.org/

McClellan, G. S.; and Larimore, J.; The changing student population. In: McClellan, G.; Stringer, J.; and Associates Eds. The Handbook of Student Affairs Administration. 3rd Ed. San Francisco: CA: Jossey-Bass; **2009,** 225–241 p.

McEwen, M. K.; and Roper, L. D.; Interracial Experiences, Knowledge, and Skills of Master's Degree Students in Graduate Programs in Student Affairs. *J. Coll. Stud. Dev.* **1994,** *35(2),* 81–87. Retrieved from http://www.jcsdonline.org/accestojcsd.html

Mueller, J. A.; and Pope, R. L.; The relationship between multicultural competence and white racial consciousness among student affairs practitioners. *J. Coll. Stud. Dev.* **2001,** *42(2),* 133–144. Retrieved from http://www.jcsdonline.org/accestojcsd.html

NASPA Student Affairs Administrators in Higher Education. Graduate program directory. Retrieved from http://www.naspa.org/career/gradprograms/search.cfm, **2012**.

Neville, H. A.; Heppner, M. J.; Louie, C. E.; Thompson, C. E.; Brooks, L.; and Baker, C. E.; The impact of multicultural training on white racial identity attitudes and therapy competencies. *Prof. Psychol.: Res. Pract.* **1996,** *27(1),* 83–89. doi: 10.1037/0735-7028.27.1.83

Pope, R. L.; and Mueller, J. A.; Faculty and curriculum: examining multicultural competence and inclusion. *J. Coll. Stud. Dev.* **2005,** *46(6),* 679–688. doi: 10.1353/csd.2005.0065

Pope, R. L.; and Mueller, J. A.; Multicultural competence. In: Student Services: Handbook for the Profession. Schuh, J. H., Jones, S. R., Harper, S. R., and Associates Eds. 5th Ed. San Francisco: CA: Jossey-Bass; **2011,** 337–352 p.

Pope, R. L.; Mueller, J. A.; and Reynolds, A. L.; Looking back and moving forward: future directions for diversity research in student affairs. *J. Coll. Stud. Dev.* **2009,** *50(6),* 640–658. doi: 10.1353/csd.0.0097

Pope, R. L.; and Reynolds, A. L.; Student affairs core competencies: integrating multicultural awareness, knowledge, and skills. *J. Coll. Stud. Dev.* **1997,** *38(3),* 266–277. Retrieved from http://www.jcsdonline.org/accestojcsd.html

Pope, R. L.; Reynolds, A. L.; and Mueller, J. A.; Multicultural Competence in Student Affairs. San Francisco, CA: Jossey-Bass; **2004**.

Priester, P. E.; Jones, J. E.; Jackson-Bailey, C. M.; Jana-Masri, A.; Jordan, E. X.; and Metz, A. J.; An analysis of content and instructional strategies in multicultural counseling courses. *J. Multicult. Coun. Dev.* **2008,** *36(1),* 29–39. Retrieved from Professional Competency Areas for Student Affairs Practitioners. A joint Publication of ACPA and NASPA. Retrieved from http://www2.myacpa.org/img/Professional_Competencies.pdf, **2010**.

Ramsey, M.; How to create a climate for cultural diversity appreciation within the classroom. In: Confronting Prejudice and Racism During Multicultural Training. Kiselica, M. S.; Ed. Alexandria, VA: American Counseling Association; **1999,** 25–42 p.

Reason, R. D.; and Broido, E. M.; Issues and strategies for social justice allies (and the student affairs professionals who hope to encourage them). *New Direct. Stud. Serv.* **2005,** *110,* 81–89. doi: 10.1002/ss.167

Reason, R. D.; and Davis, T. L.; Antecedents, precursors, and concurrent concepts in the development of social justice attitudes and actions. *New Direct. Stud. Serv.* **2005,** *110,* 5–15. doi: 10.1002/ss.161 [NOTE BOTH ENTRIES FROM REASON ARE FROM THE SAME MONOGRAPH, 110 IS THE ISSUE NUMBER]

Reynolds, A.; Helping College Students: Developing Essential Support Skills for Student Affairs Practice. San Francisco, CA: Jossey Bass; **2009**.

Reynolds, A. L.; and Rivera, L. M.; The relationship between personal characteristics, multicultural attitudes, and self-reported multicultural competence of graduate students. *Train. Educ. Prof. Psychol.* **2012,** *6(3),* 167–173. doi: 10.1037/a0029774

Shuford, B. C.; Multicultural affairs. In: Rentz's Student Affairs Practice in Higher Education. Zhang, N.; and Associates Eds. Springfield, IL: charles C. Thomas Publisher Ltd.; **2011,** 245–280 p.

Smith-Adcock, S.; Ropers-Huilman, B.; and Choate, L. H.; Feminist teaching in counselor education: promoting multicultural understanding. *J. Multicult. Coun. Dev.* **2004,** *32,* extra 402–413. Retrieved from http://www.counseling.org/publications/journals.aspx [NOTE "EXTRA" IS THE LISTING]

Sue, D. W.; Bernier, J. E.; Durran, A.; Feinberg, L.; Pedersen, P.; Smith, J.; et al. Position paper: cross-cultural counseling competencies. *The Coun. Psychol.* **1982,** *10,* 2, 45–52. Retrieved from http://tcp.sagepub.com/ doi: 10.1177/0011000082102008

Talbot, D.; Master's students' perspectives on their graduate education regarding issues of diversity. *NASPA J.* **1996b,** *33,3,* 163–178. Retrieved from http://journals.naspa.org/

Talbot, D. M.; and Kocarke, C.; Student affairs graduate faculty members' knowledge, comfort, and behaviors regarding issues of diversity. *J. Coll. Stud. Dev.* **1997,** *38(3),* 278–287. Retrieved from http://www.jcsdonline.org/accestojcsd.html

US Bureau of the Census. Statistical Abstract of the United States: PDF Version. Washington, DC: Government Printing Office. Retrieved from http://www.census.gov/compendia/statab/2010/2010edition.html, **2010**.

Waple, J. A.; An assessment of skills and competencies necessary for entry-level student affairs work. *NASPA J.* **2006,** *43(1),* 1–18. Retrieved from http://journals.naspa.org/

CHAPTER 10

BEYOND TOLERANCE: DEVELOPING MULTICULTURAL COMPETENCIES IN PRE-SERVICE TEACHERS

GEORGE STILL and MAUREEN SQUIRES

CONTENTS

10.1 INTRODUCTION

One of the enduring critiques of P–12 education is the advantage afforded to certain groups of both students and teachers, based on their race/ethnicity, English as their primary language, gender, ability, or socioeconomic status. Further, policies designed to provide additional opportunities for marginalized groups use an "assimilation" mindset. The primary goal of this "assimilation" mindset is to create a system where minorities must become indoctrinated into the culture and values of the dominant group. In this system, the dominant group defines measures of success. There is little, if any, consideration for the merit of minority experiences.

Because social and ethnic minorities experience systemic, compounding inequalities (Kozol, 1991) in education, multicultural competencies are the set of skills and dispositions designed to integrate people of color and other marginalized groups into mainstream society (Baltodano, 2006). The intention is to disrupt the assimilation model that has been a part of all social institutions and further the agenda for a pluralistic society. The goal of multicultural education in any context must be to value the diversity of students, address faculty members' own personal biases, and help students increase their multicultural competencies. This is specifically important for pre-service teachers. Teacher education programs must prepare educators to change the P–12 education system that has for so long advantaged certain populations.

10.2 LITERATURE REVIEW

According to the US Census Bureau (2012), the United States will continue to become considerably older and more racially and ethnically diverse over the next 40 years. By 2043, the current "majority" will no longer exist; "minorities" will comprise more than half of the population. By 2060, the nation's population of children is projected to be 67 per cent minority. Most of these children will attend public schools where their ethnic heritage is not proportionally represented in the demographics of the teaching force. That is because teaching remains an "overwhelmingly female" (84%) and "strikingly White" (84%) profession (Feistritzer, Griffin, and Linnajarvi, 2011, p. x). Such disproportionality can negatively affect minority students' learning.

When there is a close fit between school culture and students' home culture, teaching is relevant and promotes student engagement and understanding. This close relationship is known by many phrases: "culturally relevant" (Ladson-Billings, 1995), "culturally appropriate" (Au and Mason, 1981), "culturally congruent" (Mohatt and Erickson, 1981), "culturally responsive" (Cazden, 2001), or "culturally synchronous" (Irvine, 2003). Murrell (as cited in Sleeter, 2008) adds specificity to our understanding. He uses the term "community teacher" to describe the teacher who "possesses the contextualized knowledge of the culture, community, and identity of the children and families he or she serves and draws on his knowledge to create the core teaching practices necessary for effectiveness in diverse settings" (p. 215). When there is a loose fit between school culture and students' home culture, teaching estranges groups of students by disregarding minority culture and reinforcing mainstream culture. Such a mismatch results in misunderstandings between teachers and students and impedes students' learning (Ambe, 2006; Delpit, 2006; Ladson-Billings, 1995).

10.2.1 CULTURAL INEQUITY

In addition to misunderstandings, research indicates that some teachers hold deficit-based perceptions of marginalized children (Ambe, 2006; Davis, 2009; Sleeter, 2008). Deficit-oriented teachers believe that children of color lack the knowledge and skills, sometimes referred to as cultural capital, to be successful in school. Proponents see the school's responsibility as rescuing (often alienating) "culturally deprived" children from their own heritage (Banks, 2007; 2008).

By contrast, the cultural difference paradigm sees children of diverse backgrounds as different, not deficient. Difference-oriented teachers believe that diverse cultures are full and enriching. Their role is to change school and classroom structures to reflect diverse cultures. They draw upon multiple languages, participation and communication styles, and worldviews to effectively reach and teach their students. For instance, it was found from Banks' (2007) discussion of studies conducted in Native American classrooms that students were more engaged in classrooms that used visual learning, local talk patterns, and bilingual approach strategies that honor the students' home culture. If all students are to achieve their greatest potential, it is imperative that teachers be multiculturally competent. Individual and collective work toward this goal can begin during teacher preparation.

10.2.2 TEACHERS' (NEED FOR) MULTICULTURAL COMPETENCE

Research indicates that although teachers generally are competent in their content area, they often overlook or lack the skills and dispositions necessary to successfully teach diverse students (Benton-Borghi and Chang, 2012; Sogunro, 2001). Banks (2008) argues that effective teacher education programs are "essential for the successful implementation of multicultural education in the schools" (p. 117). Benton-Borghi and Chang studied pre-service and in-service teachers' sense of efficacy to teach diverse student populations. They found that both groups felt less efficacious to teach students of color and students with special needs than nondiverse students. Benton-Borghi and Chang focused on teachers' sense of efficacy, in part, because numerous studies have indicated that it is a "reliable and valid predictor of student achievement" (p. 36). Here, teaching practice is a mediating factor, connecting efficacy and achievement. This supports similar research conducted by Darling-Hammond, Chung, and Frelow (2002), which found that a sense of preparedness was the second strongest predictor of teachers' sense of efficacy. If teachers feel unprepared, they are less secure in their efficaciousness to teach diverse students and are less likely to effectively teach this population.

Baltodano (2006) discussed a way that teacher education programs and accreditation processes attempt to include multiculturalism as part of the pre-teacher education experience as found in National Council for Accreditation of Teacher Education (NCATE). She asserted that a cursory process was insufficient. Attempting to include information about lesson plans and courses about linguistic and ethnic diversity could help pre-service teachers learn about the "other." However, the curriculum does not demand that pre-service teachers internalize multiculturalism. As discussed in the works of R.W. Connell (1995) and Gloria Anzaldúa (2007), learning about the other does not address the role of power and dominance, especially within the teaching context, or the role teachers have played in sustaining inequality. The cursory use of multiculturalism often amounts to a degree of "cultural tourism" on the part of dominant groups, where they are able to travel to learn about marginalized groups and return safely to their dominant position without ever having to challenge their own biases or the institutional process.

Also known as "culture on wheels," in P–12 schools, this tourist approach includes celebrating cultures by month (i.e. "Black History Month," "Latin America Month," etc.) via food, art, music, or dance festivals (Ambe, 2006). In college classrooms, cultural tourism can be exemplified by a single multiculturalism course. This approach minimizes marginalized groups by treating their heritage as additive, something studied a few days of the school year or during a semester. Such one-shot deals typically are not successful in significantly changing one's attitudes and actions in pluralistic contexts. For example, Lobb (2012) studied five college students taking an elective multicultural education course. Considering participant interviews, weekly reflection papers, and other written work, she found that college students moved somewhat toward greater self-awareness of multicultural topics throughout the semester. Yet, there was no indication that these minor changes persisted beyond the semester or were systemic. Similarly, Gomez (1993) found that a single course produced limited to moderate changes in pre-service teachers' beliefs about diversity. Multiple factors such as motivation, commitment to change, course curriculum, attitudes of faculty, etc., have been shown to affect college student development, and multicultural competencies are no exception.

It is our opinion that future teachers need a deeper level of understanding beyond a perfunctory look at multiculturalism or an approach that relegates diversity to the "other." We see the importance of understanding the role of self in this process. As Anzaldua proposes in *Borderlands*, there is a need to understand the role of self and create a new space that values diversity, what she refers to as a *mestiza* consciousness (2006). This new consciousness would value ambiguity and further respect difference and understand the "borders" that we cross. Embedded, systemic approaches in which pre-service teachers continually reflect on and critically think about their underlying assumptions are effective at transforming dispositions (Nieto, 2007). Yet often multicultural components are not consistently woven throughout curricula in higher education institutions (Ambe, 2006).

10.2.3 REFLECTION AND FIELD EXPERIENCE

One area worthy of specific mention is field experience. As teacher education programs continue to develop additional opportunities for clinical experience, they need to incorporate activities that augment multicultural competencies. College students often do not make the connections between what is learned in their courses to their field experience. Multicultural competencies are no exception.

Extant literature (Ambe, 2006; He and Cooper, 2009; Lobb, 2012) highlights the importance of reflection and field experience in developing pre-service teachers' multicultural competency. To effectively teach diverse students, pre-service teachers must first critically examine their own experiences and underlying assumptions about diversity. Pre-service teachers also need to learn about their students' cultures and critique their instructional practices (rooted in their beliefs) in light of all students. As Cochran-Smith (1995) explains, prospective teachers

> need opportunities to examine much of what is usually unexamined in the tightly braided relationships of language, culture, and power in schools and schooling. This kind of examination inevitably begins with our own histories as human beings and educators—our own cultural, racial, and linguistic backgrounds and our own experiences as raced, classed, and gendered children, parents, and teachers in the world. It also includes a close look at the tacit assumptions we make about the motivations and behaviors of other children, other parents, and other teachers and about the pedagogies we deem most appropriate for learners who are like us and who are not like us. (p. 500)

Teacher preparation programs can support pre-service teachers in becoming reflective practitioners through program and course design. This can include providing scaffolding questions and modeling difficult discussions. In addition, student teaching provides pre-service teachers the opportunity to put theory into practice. Ideally, pre-service teachers would be placed with cooperating teachers who use culturally congruent practices (Benton-Borghi and Chang, 2012; Delano-Oriaran, 2012). In that context, they can observe and practice critical pedagogy.

A cross-culturally competent person is one who is committed to continually examining self and is open to personal growth, intellectually and emotionally, who supports the unity of human beings and appreciates the differences between people of different cultures and who is dedicated to fighting racism on all levels and in all forms (Gudykunst and Kim,1984), and Bennett (1995), as cited in McAllister and Irvine, 2000). Such commitment is not evident in all people. In reality, many White people resist, consciously or unconsciously, self-reflection and accountability and reject counterstories. This lack of critical analysis and awareness surfaces in colorblind behaviors and beliefs.

10.2.4 CONDITIONS THAT SUSTAIN WHITE PRIVILEGE

Bonilla-Silva (2010) writes about "new racism" or "contemporary racial inequality [reproduced through] practices that are subtle, institutional, and apparently nonracial" (p. 3). One of these practices is the adoption of colorblind ideology. On the surface, colorblindness appears to many as nonracist or antidiscrimination (Karafantis, Pierre-Louis, and Lewandowski, 2010; Thompson, 1999). It attempts to achieve social harmony by disregarding or ignoring focus on group differences on the basis of race, ethnicity, or culture (Karafantis et al., 2010). From this perspective, eliminating difference, particularly based on color, allows for equal treatment of all people. Yet, this is a misunderstanding. On closer examination, colorblindness is revealed as a means to dismiss difference, thereby maintaining the status quo, which is advantageous to certain groups while disadvantageous to others. As Carr argued, "colorblindness is not the opposite of racism, it is another form of racism" (as cited in Karafantis et al., 2010, p. 691).

As described by Gordon (2005b), colorblindness is a "self-imposed blindness, a refusal [not an inability] to see race and acknowledge White privilege" (p. 281). It exists at both individual and systemic levels, in multiple realms (i.e. political, economic, legal, and educational) throughout the society. Through colorblind ideology, Whites blame minorities for their current position in life, attributing poverty, low employment rates, or residential segregation to work ethic, cultural deficiencies, or natural phenomena (Bonilla-Silva, 2010). From such a hegemonic lens, they often disregard the gap that still exists between themselves and minorities, believing that a state of sameness is ideal and nearly has been achieved. In reality, today, minorities have significantly higher infant mortality rates and school dropout rates than nonminorities; they also earn less income, have fewer assets, attain less education, and have shorter life expectancy (Delgado and Stefancic, 2001).

In the field of education, colorblindness presents in several ways. For instance, a practicing teacher states thus: "I don't see color; I just see children." Or a student teacher uses literature and research exclusively from mainstream authors and scholars to avoid dealing with controversial texts. People may adopt colorblind ideologies, believing that they can "overcome and 'conquer it [racism and discrimination] by ignoring it' " (Lorde (1982), as cited in Milner, 2007, p. 392). However, rather than rectifying the situation, such denials continue to harm minorities. First, by denying race, White teachers do not have to acknowledge themselves as racialized beings or examine their race-based beliefs and actions. Without critical reflection, they are left in the dark, unable to work toward a democratic classroom or society. Second, minority students' cultures are dismissed as worthless, and the obstacles they face are erased. Their true identities, often marked by struggle and disparity, are replaced by sterilized versions.

From another perspective, people elect to be colorblind because they can then claim ignorance and innocence of involvement in systems that perpetuate racial inequality (Gordon, 2005a; Thompson, 1999). A social consciousness of sameness calls individuals who question difference based on race, gender, racists, etc. Therefore, not to be prejudiced means pretending not to see skin color. That is, if the phenomenon does not exist, then a person cannot be its participant. Such an argument attempts to negate Whites' complicity in racism, yet actually highlights attempts to dodge accountability. As Thompson (1998) writes, "There is no such thing as racial innocence; there is only racial responsibility or irresponsibility" (p. 524).

People also espouse colorblindness as it allows them to maintain their power and racial privilege (Bonilla-Silva, 2010; Delgado and Stefancic, 2001; Thompson, 1999). By claiming not to see skin color, Whites can preserve the status quo, whereby they hold primary positions of status and authority, whereas minorities are relegated to secondary positions. Seeing color differences means acknowledging oppression. In unpacking her own position, McIntosh (1989) came to understand White privilege as an

> elusive and fugitive subject. The pressure to avoid it is great, for in facing it I must give up the myth of meritocracy. If these things are true, this is not such a free country; one's life is not what one makes it; many doors open for certain people through no virtues of heir own. (n.p.)

Regardless of intent, colorblindness is detrimental to society as a whole. It is often hidden from view and has the ability to seep into and become deeply entrenched in all facets of society, including education. It becomes an unquestioned "norm."

Seeing color is important. Critical race theorists (CRTs) believe "Only aggressive, color-conscious efforts to change the way things are will do much to ameliorate misery" of marginalized people (Delgado and Stefancic, 2001, p. 22). Teacher education preparation programs need to find ways of incorporating content that will address colorblind marginalization. One of the most important ways of ameliorating this issue will be to integrate discussions of inequality throughout coursework.

10.3 METHOD

CRT comprises five tenets that apply to educational research. First, it posits that racism is endemic and deeply ingrained in American life. Second, CRT challenges claims of neutrality, objectivity, colorblindness, and meritocracy. Third, CRT is devoted to social justice. Fourth, CRT gives voice to people who have historically been silenced. Last, CRT is necessarily interdisciplinary (Evans-Winters and Hoff, 2011; Ladson-Billings and Tate, 1995). CRT provides the framework for this study.

The teacher education program studied exists within a division in a small, liberal arts institution with approximately 6,000 students. The teacher education program currently offers undergraduate and graduate degrees in childhood education (1–6), special education (B–12), literacy (B–12), adolescence, and education (7–12). At the time of the survey, there were 367 pre-service teachers enrolled in teacher education preparation programs at the college. Of these pre-service teachers, 77 per cent were women and 23 per cent were men. In addition, 88 per cent of pre-service teachers were White and 12 per cent from minority populations.

The authors of this chapter have examined the role of coursework in developing pre-service teachers' multicultural competencies. The program faculty define multicultural competency as follows: "Our graduates adapt their planning and instruction to meet the diverse needs and learning styles of students of all backgrounds and abilities; they strive for instructional excellence and develop multicultural competencies that inform all that they do" (Ackland, Schnackenberg, Simard, and Still, 2013). The emphasis in this context rests primarily in a social justice paradigm, in which the goal is to challenge the institution and process of schooling to create a space for diversity and inclusion of all students. Pre-service teachers are expected to have a critical sense of self and the role they play in the education process, particularly as it relates to power in the P–12 education. They further should advocate for the inclusion of all students. Finally, they should constantly evaluate their own biases.

The survey was administered in the Fall 2011 semester to all entering pre-service teachers in the following programs: BS in childhood education, combined BS in childhood education and MSEd in special education, BA/MST adolescence education program, and MST childhood/adolescence education. The same survey was also administered to pre-service teachers post coursework in multicultural competencies in the following programs in Spring 2012, Fall 2012, and Spring 2013: BS in childhood education, combined BS in childhood education and MSEd in special education, BA/MST adolescence education, and MST childhood/adolescence Education, MSEd in special education, and MSEd in literacy education. For the MST, MSEd, and BA/MST programs, this was a longitudinal look at their development through coursework. However, the BS and BS/MSEd will complete their pre-student teaching coursework in Fall 2014, and a follow-up administration is planned for those pre-service teachers at that time.

We, the researchers, were interested in exploring pre-service teachers' personal awareness of multicultural competency and ability to apply this awareness to a classroom context. In both an entering pre-service teacher survey and post-coursework survey, respondents were asked two open-ended questions: (1) What does being a multiculturally competent and community-focused educator mean to you? (2) What would you do in a classroom to demonstrate your multicultural competencies? In addition, one Likert scale question was asked: Do you consider yourself multiculturally competent? The two open-ended questions were designed to assess what pre-service teachers think about multicultural competencies and what this would look like in a classroom. Responses to the two open-ended responses could then be compared with respondents' self-assessments of their multicultural competencies. The goal for the survey was to provide a baseline for the multicultural competencies of pre-service teachers who enter the program. In addition, it measured the success of the program in augmenting pre-service teachers' multicultural competencies.

The survey was administered in courses in the various degree programs. The entering student survey was given in several of the first education courses that students take in their degree programs. None of these entry-level classes focuses strongly on multicultural competencies. The post-coursework administration occurred after the specific courses where multicultural competencies are a primary focus: in EDU 130 for the BS and BS/MSED programs or EDR 514 for all other programs. The instrument was given during class, in both paper and electronic format depending on the preference of the course instructor. The electronic administration was conducted using computer labs that helped to stabilize the response rate between administration formats.

10.4 FINDINGS AND ANALYSIS

Survey respondents were almost exclusively White, which provides a wealth of information regarding pre-service teachers' thinking from the perspective of racial privilege. Of pre-service teachers who identified their race, there were 130 White respondents, three Black, one Hispanic, and one Asian. In addition, 38 respondents, of the 104 total respondents, were male, which is fairly representative of the gender stratification of teachers. Both of these dynamics are highly representative of the overall teaching force. It also exposes one of the problems in the profession right now in terms of the increased diversity in the student population compared with the teaching workforce (Irvine, 2003).

In terms of pre-service teachers' self-assessments of their multicultural competencies, respondents to the survey felt that they were multiculturally competent. The survey response options were *very competent*, *somewhat competent*, *not very competent*, and *not at all competent*. Of the 63 entering pre-service teacher responses, 25.4 per cent of respondents reported that they were *very competent* and 74.6 per cent reported that they were *somewhat competent*. No respondents selected *not very competent* and *not at all competent* responses. The responses shifted once pre-service teachers took coursework in multicultural competencies. Of the 94 responses, 30.7 per cent of respondents saw themselves as *very competent*, 68 per cent *somewhat competent*, and 1.3 per cent *not very competent*.

Likert ratings were discrepant from open-ended responses. Many respondents rated themselves higher in multicultural competence than their open-ended responses about multiculturalism in the classroom justified. The analysis revealed pre-service teachers' responses in general were in the following themes: minority deficit perceptions, treating minorities as "others" through a cultural tourist approach, and color and class blindness. Once pre-service teachers had completed EDR 514[1] or EDU 130[2] in multicultural competencies, a few responses rose to a level of cultural appropriateness that a pluralistic, social justice/equity approach represents; however, many of the responses remained in a lower level of understanding. Each of these themes is described below.

10.4.1 DEFICIT-BASED PERCEPTIONS

The deficit-based perspective is that students in the P–12 classroom lack skills based on their identity and that the role of the teacher is to ameliorate such deficits. There were no overt deficit-based responses by pre-service teachers in our survey administrations. However, there are several subtle ways given below that deficit-based perceptions could be found in the survey.

Several respondents saw the need to compensate for difference by attempting to bolster the "inadequate" skills of diverse students or provide common (hegemonic) experiences in the classroom. In terms of pre-coursework responses, one respondent stated that multiculturalism is "being able to adapt to situations and different cultures and working with needs based on where you're teaching and where the students are from." The response used two phrases that show a deficit-based perception, which were "needs" and "where the students are from." Both phrases

---

1 The course description states the following as the course objectives for EDR 514: affirm; respect; and value cultural, linguistic, and ethnic diversity through the development of multicultural competencies. Respond to increasing diversity in classrooms through conscious awareness and use of multiple, integrated, practical, and creative problem-solving strategies adapted to the specific needs and interests of children in all areas of literacy. Explore literature that reflects diversity and issues of social justice. Study systems of dominance and marginalization with specific attention to the role of literacy within such systems.

2 The course description states the following as the course objectives for EDU 130: investigating; thinking critically; and reflecting on ethics, relationships, and multicultural competencies in education. Exploration of the ethical and practical dimensions of teaching within the diverse contexts that teachers and their students bring to the classroom community. Students will be expected to develop an understanding of multicultural teaching competencies and engage in self-reflection to identify and monitor their strengths and areas in need of improvement.

presume that the differences in the child's background are deficits rather than assets to that child/community.

Another pre-service teacher responded similarly. In terms of what multicultural competencies meant, the response included language about "compensating" for difference and how this would look in a classroom. The respondent wrote about Christmas and stated "it feels good when classmates can celebrate and be around their peers during the holidays." This response also articulates the hierarchy of culture, whereby some traditions and values are celebrated while others are ignored. When attention is paid to one culture at the expense of other cultures, a message is sent: only certain traditions "count." Moreover, it is important to be a member of, or at least participate in, the endorsed culture.

In post-coursework responses, a respondent stated, "being multiculturally competent means understanding the struggles people have been through and helping them better their live(s)." This response shows the subtle way that the deficit model is part of an individual's understanding. There was no mention in the response of strengths that minorities possess. In addition, the response has language that blames people for their circumstances. The response presents the minority person experiencing inequality as lacking something that the teacher can help augment/change.

There was a similar response by another post coursework pre-service teacher. The response also blames students for their conditions, stating "I will take in the fact that not all students are equal in their ability to get a good education. There are just some factors that get in the way of achievement." In the process of displacing blame, the pre-service teacher removes responsibility from him-/her-self. Therefore, lack of academic achievement resides in the student, not the teacher. That suggests that what the student brings to the classroom determines his/her success, not the teaching or learning environment created in the classroom.

10.4.2 CULTURAL TOURISM AND THE "OTHER"

Through cultural tourism, pre-service teachers saw the role of multicultural education as exposing students to difference through events, holidays, etc., without altering their "curriculum" or creating a more inclusive space for all students. In the entering pre-service teacher survey, respondents said the role of education was to provide space for minority students to "share" about their own experiences through cultural fairs, Black History Month, and exposure to holidays of non-Christian groups. The majority of the class curriculum remained focused on the institutionally sanctioned activities of the dominant group. There were only four responses in the entering pre-service teacher survey of the 63 total that went beyond tourism. That is, they moved beyond simply "traveling" to learn about other groups to confronting personal and institutional processes that maintain privilege.

The cultural tourist approach is best exemplified through the following type of response: "I would ask students to each bring in information about their culture and ask them to share." A similar response stated: "[I would] explain to students all of the different cultures throughout the world. I would set up an international week of some sort where the children would come in and share the different places they're from and the different things they do at home that pertain to their culture." In both the actions of the teacher and students, the classroom appears to be set up for students to take a trip to another culture and return to "normal" programming once the events are finished. Responses such as this suggest a lack of consistent focus on diversity. Further, no focus on social justice is indicated. The work of students is mainly focused on minority students sharing about their own experiences.

In other responses entering pre-service teachers describe how they would help students learn by "encouraging parents of different cultures to come into my classroom and present to the students their culture in hopes that students will come to understand why a certain student might act a little differently or dress in a different manner." The previous two responses suggest that the work is placed either directly on the students or on the parents. This is indicative of one tenet of CRT, which asserts that racism is systemic in American life. Due to difference, "the other" is deficient and therefore must attempt to explain or justify him-/her-self. The student is further marginalized by drawing attention to his/her difference. In addition, it essentializes the experiences of an entire group of people into a single person. Just like a tourist in a foreign context, the lifestyle of an entire group of people is categorized based on the tourist's experience on that vacation.

Often holidays are where this will occur as exemplified through the following statement: "I would make sure other culture's holidays and traditions are noticed and taught to all students" in his/her response about how multiculturalism would be implemented in the classroom. The respondent seemed only to consider holidays as a space for difference and used the word "noticed" rather than "honored," "incorporated," or even "included."

Some responses also exemplified the role of the teacher in this context as the tour guide. One respondent stated thus: "In a classroom, I would have a lesson on different cultures for different days of the week." Another respondent stated thus: "I'd pose questions, talk about where I'm from, my heritage, and my own culture and try and relate to students and ask them to relate to each other." Or this type of response: "I would first research the culture, then I would set lesson plans up for the children around what I have learned. I would bring in books to read about other cultures. They could do painting and sing songs, etc. Just let the children have a grasp on what that culture is about." These responses show that multiculturalism is teacher-led but still just a tourist approach because it is not incorporated throughout the curriculum. These responses show the emphasis on activities, which are focused primarily on student exposure to topics rather than emphasizing in all aspects of their instruction. In addition, it also removes students from engaging in inquiry because it is important for students to do personal/individual work. In the above-mentioned response, the teacher is the primary person researching other cultures.

A big part of the teacher tour guide modality was the teacher asking students to do research. Many respondents used this as part of their approach to demonstrate their multicultural competencies. "I would do around the world projects where the students take different cultures and learn about them." Or "I would have the children do projects where it would incorporate cultures that they maybe didn't even know existed." While on the surface, additional knowledge is important and students in the classroom should play an active role in learning about diversity, research projects expand intellectual knowledge but rarely change behaviors and the subtle forms of inequality.

Additionally, well-intentioned research projects can have inadvertent negative consequences. A respondent stated: "I would have the children research a famous African American and then research a famous Caucasian from the civil war. They will see how it will be much harder to find information on the African American and I will show them how unfair that is." This response takes students on a tour through researching race. The pre-service teacher attempts to use research of minorities as a means of exposing inequality and social justice but still only scratches the surface. Moreover, this type of project could reinforce an understanding that minorities do not contribute to society. In addition, the teacher is still playing a major role in the learning process through "showing" students how unfair this is because children and adolescents can struggle with internalizing external events. Often, pre-service teachers will not naturally connect historical events to present inequalities.

Some responses suggest the process of cultural tourism when pre-service teachers dichotomize groups as "like me" or "other than me." One respondent stated thus: "Learn about different countries and the different ways they do things compared to what we do here." Another respondent stated thus: "I would teach about the different cultures and how they are important in everyday life. All students would be exposed to different cultures." Both respondents focused on the differences between cultures. This mindset perpetuates stereotypes. In the process of learning about other cultures, students are not exposed to their own privilege. In addition, students often become disconnected from the experiences in other contexts because they do not make the connections to other topics they are learning in class. In these examples, multicultural education is a peripheral activity.

Respondents who had completed coursework in multicultural competencies demonstrated a deeper understanding of multicultural competencies than entering pre-service teachers. The responses were less likely to include language of cultural tourism. Fifteen of 74 total responses contained language of cultural tourism. This compares to 59 of 63 total responses of entering pre-service teachers, which included language of cultural tourism.

There were still respondents who thought incorporating multicultural competencies in the classroom meant having students learn about diversity through projects. This was present in this response: I would "have students teach each other about their family's heritage." Another respondent stated, "There are many activities and ideas to share with my students. I could set up projects in which the students are able to learn more about other people and the differences that they have." Another respondent stated, "I would have a lesson plan where I have every student in the classroom teach each other something about their culture." In all of these responses, pre-service teachers treat culture as additive, not integral and integrated throughout the curriculum. Moreover, they neglect to discuss related issues, such as inequality and social justice. Neither social justice nor inequality is mentioned in the previous examples. Such responses suggest that pre-service teachers address multiculturalism at a basic level. That is, they direct students to look outward at other cultures without looking inward to challenge personal assumptions, biases, or privilege.

In addition, some post-coursework responses portray the teacher as the tour guide. One respondent said, "I would demonstrate multicultural competency by having activities and lessons that broaden the views of students. I would expose them to people and groups different than themselves through these." Another respondent stated thus: "I would try to integrate learning about different cultures in my weekly lesson plans. Once a month (or twice a month depending on my schedule) I would choose a particular culture and dress in their traditional clothing and spend half of the day learning and teaching the children about that culture."

Another respondent discussed that he/she would, "do different activities with books, lessons, movies, holidays, and guest speakers. I would teach students multiculturalism." In these responses, the teacher is primarily responsible for curriculum, where multiculturalism is not woven throughout the curriculum. It is as if the curriculum is constructed of distinct spokes of a wheel (i.e. literacy, math, social studies, science, art, music, physical education, multiculturalism, etc.), rather than connect by a multicultural cog.

Another challenge is demonstrated in the link between pre-service teachers' current perspectives about multicultural competencies and their own P–12 experience. One respondent recalled in her schooling experience, "my first grade teacher set aside a day where students brought in food that represented their culture. Reflecting on that now, I see why she did it." In this response, the way to increase multiculturalism in the classroom is by learning about the other.

10.4.3 COLORBLIND PERCEPTIONS

The other theme that was present in survey responses was based on colorblind ideologies. Colorblind ideologies are more subtle, institutional, and apparently nonracial (Bonilla-Silva, 2010). They are a means to dismiss difference (or inequality), thereby maintaining the status quo, which advantages certain groups while disadvantaging others.

There was a strong connection between cultural tourism, the concept of the "other" and colorblind responses. These three concepts attempt to minimize the importance of difference, inequality, hegemonic masculinity, and dominance in social systems. In the entering pre-service teacher responses, colorblind ideology could be found through the lack of responses that discussed difference beyond ethnicity. Pre-service teachers primarily thought that multicultural competencies meant attending to cultural/ethnic differences. There were only a few responses that discussed race, gender, or class inequalities.

Another way that this response occurred in the entering pre-service teacher survey was through responses that whitewashed difference. One respondent discussed the United States as a melting pot, which is based on an assimilation model where minorities are expected to shed their heritage for dominant ideologies. Respondents also used terms such as "normal" and "tolerance" when describing difference in the classroom. These statements, from a privileged perspective, may not appear discriminatory, but they neither incorporate diversity nor attempt to enact social change in the classroom. In addition, "equally" and "fairly" consistent were present in responses. Both of these statements are based on the premise that students have equal access to resources and the same institutional experience in society. Both terms show a naivety on the part of entering pre-service teachers by assuming that treating all students the same is equivalent to treating all students fairly and equally. In reality, treating all students the same disregards their unique talents and needs, thereby making teaching-learning one-size-fits-all. Such standardized educational approaches fail to provide specialized instruction (designed for the individual, based on prior knowledge, and scaffolded to appropriately challenge and support each student).

There were other more overtly colorblind responses by new pre-service teachers. One respondent stated that multicultural competencies consist of "allowing other cultures in the classroom and making them all acceptable." In this response, the teacher is the gatekeeper and determines what is acceptable in the classroom. The term "allowing" does not establish a system where pluralism and social justice are present because it still relies on traditional notions of power and authority. Another pre-service teacher responded that multiculturalism is "that I let the students learn that we are all one no matter the color." This response shows the emphasis on unity without any consideration of the history of inequality and how it continues to impact society today. The denial of inequality and emphasis on unity or similarity is part of the colorblind ideology.

In the post-coursework responses, the emphasis on similarity/unity was a consistent challenge. A respondent commented: I would "treat everyone equally." Just like in the entering pre-service teacher survey, it assumes that equality and fairness means the same; but there are differences. This challenge was more direct in the following response: "No matter where anyone is from, what language they speak, what disabilities, financial status, ethnicity, sex, or sexual orientation they may have, a good educator will not base his/her approach on those identities." This response highlights the minimization of difference that occurs in colorblind ideology. Instead of using students' identity to mold teaching, the respondent decides to treat everyone the same.

Another challenge can be seen in a response that normalized dominant behavior. One respondent stated thus: "They may not always be like me or their peers, but their situation needs to be given value and understanding." This response highlights the tendency for individuals that experience privilege to establish the parameters for what defines difference.

10.4.4 PLURALISM AND SOCIAL JUSTICE

Multicultural and inclusive education should include the following components: educators who reflect and address their personal biases, pluralistic curriculum with an emphasis on social justice, and a classroom free of peer discriminatory practices. The survey demonstrated that respondents articulated the above-mentioned characteristics in their perceptions of what multicultural competencies are and how it would look in the classroom. There also was evidence that the program played a role in the development of pre-service teachers' multicultural competencies.

There were a few entering pre-service teacher responses that demonstrated a higher-level understanding of multicultural competencies and how these should impact the classroom. Although these responses were not inclusive of all three characteristics, social justice and pluralism were included. One specific area that entering pre-service teachers struggled in understanding was multiculturalism as woven throughout the curriculum.

One example from an entering pre-service teacher is "I would need to let everyone know that there is no prejudice allowed in the classroom, and that they should eliminate it outside of the classroom as well." This statement shows a respondent with a strong understanding that prejudice is something that should not be present in the classroom but no mention of his/her own bias and also no mention about curriculum. A similar comment by another pre-service teacher follows: “If there is prejudice in the class the teacher needs to figure out a creative way to teach the children that it is not right and why. At a young age children’s opinions are being formed and teaching them early about multicultural differences can help them be more educated but also more accepting as they grow older.” This response also shows a basic understanding of inequality, but minimal understanding of how to address it.

Another pre-service teacher demonstrated a deeper understanding of multiculturalism. The respondent noted, “promoting empathy on a number of different social, political and historical topics.” In the same response to implementation the respondent stated: “I would encourage students to test their boundaries and values against real world situations, get in the mix and learn more from firsthand experience whether or not your biases and prejudgments are true.” In this example, the pre-service teacher showed an understanding that multiculturalism should be incorporated as part of the curriculum, include social justice components, and address student bias. The major component missing was self-assessment.

The post coursework responses demonstrated the impact of the curriculum on pre-service teachers’ perceptions. Some responses were inclusive of the above-mentioned characteristics. Respondents began to make the connection that addressing personal and student bias and designing an inclusive curriculum with social justice components were necessary to educate all students.

One pre-service teacher stated multiculturalism in the classroom would “address stereotype statements, and inform students that say these things what they are actually doing. Making sure to not make general statements toward one group, but include everyone.” In the same respondent’s definition he/she stated thus: “multicultural competency is to be aware of different cultures and understand the treatment and oppression that some cultures face. In addition, it is to understand the target and dominant groups in a community and inform others about the inequality.” The response by this pre-service teacher shows multicultural education is incorporated into the curriculum, addresses bias, and works for social change. The only component missing was mention of dealing with his/her own personal bias. However, in general, this represented a more robust and thoughtful response concerning multicultural competencies.

Another respondent commented on how multicultural competencies would look in a classroom: "I would make sure I get to know each student as well as their families. I would also make sure my classroom books and tools that the students use every day show a diverse population. This will allow students to see themselves in their classroom and be able to feel like they belong. I believe that doing this will encourage students to be active and proud of their schoolwork, because they will see the connections to their own life, background and cultures." This response shows that knowing students and caring for them, diversifying the materials, and making the curriculum relevant would help educate all students.

One of the respondents who demonstrated the strongest understanding wrote the following: "Being a multiculturally competent and community-focused educator means a couple different things. I think it means having the knowledge of different cultures. I think it means knowing about privilege, oppression, dominance, discrimination and stereotypes. I think it means not specifically allowing the past aspects in the classroom. Equality and equal opportunity in the classroom is essential." In the same response about incorporating it into the curriculum, a pre-service teacher stated thus: "I would make sure everything was planned with equality and used equal opportunity to demonstrate my multicultural competencies. I would incorporate all cultures into my curriculum. I would not use materials that only display the dominant groups. I would give all my students equal opportunity also." This demonstrated all of the components of a multiculturally competent educator. There was specific attention paid to incorporating multiculturalism throughout the curriculum, examining dominance and oppression, developing personal awareness, and emphasizing social justice.

10.5 DISCUSSION AND IMPLICATIONS

Teaching as a profession continues to be a primarily White, female occupation and the teacher education program sampled through this survey was no exception. The results demonstrated some of the challenges present with a primarily White and therefore privileged pre-service teacher population. Pre-service teachers entering the program had minimal understanding about what multicultural competencies were and how to incorporate them in the classroom.

Respondents primarily demonstrated a tolerance and awareness-based understanding. This understanding may be in part based on their monocultural experience in P–12 education (Nieto, 2007). The monocultural perspective is one where only one cultural perspective is taught. Respondents were overwhelmingly White and middle-class, having attended elementary and secondary schools with comparable demographics. Although there was not a specific question that addressed this topic, the responses by entering pre-service teachers demonstrated that diversity was not a large part of their education, if any. Future research will examine further the P–12 experience of pre-service teachers in the program. In addition, this suggests a need for more field experiences in diverse classrooms. When educational experiences are monocultural, pre-service teachers often develop limited frameworks for conceptualizing diversity. In turn, they have limited pedagogical tools for teaching diverse students. That is not to say teachers develop and demonstrate multicultural competencies exclusively or automatically through exposure to diverse educational settings. Yet placing pre-service teachers in diverse fieldwork settings can provide experiences they previously lacked. Ideally, pre-service teachers would be paired with cooperating teachers who are multiculturally competent and serve as models and motivators to praxis. As previous research (Benton-Borghi and Chang, 2012; Delano-Oriaran, 2012) indicates, field experience and reflection influence teachers' multicultural competence.

It is not enough to simply be in a diverse setting. Pre-service teachers must also reflect on their own beliefs and underlying assumptions and how those thoughts affect their pedagogy. Several respondents commented that former teachers and school experiences directly influenced their multicultural perceptions. A few even referenced events and activities their teachers did as examples for their own actions. One of the major challenges for college faculty is working with pre-service teachers to shift their thinking about what a multicultural experience should look like because often it is at odds with their own personal experience. In order to change future students' experiences in P–12 education, it is important to alter the perceptions of pre-service teachers and increase their multicultural competencies. This can be done by problematizing situations, giving voice to historically silenced people, guiding pre-service teachers through meaningful self-reflection, and discussing and implementing explicit actions to counter prejudice.

Tolerance and awareness both were consistent themes found in the survey responses. Of the 130 total responses, awareness occurred in 45 individual responses and tolerance occurred in 10 responses. Awareness was present in 35 per cent of all responses and demonstrates one of the main challenges in changing pre-service teachers' perceptions around issues of multicultural competencies. In both the pre- and post-surveys, pre-service teachers continued to feel that being aware of difference equated to being multicultural competent. The challenge with this perception is that being aware of inequality does not remove the possibility for acting in a racist, classist, ethnocentric, or ablest manner (among other forms of inequality). It also removes pre-service teachers from their own responsibility in creating/sustaining situations of inequality because they are not thinking about their own actions beyond awareness. It is one thing to understand how "bad" someone else's life is, but it is another to actually do something about it.

Another trend was that cultural tourism consistently occurred in entering pre-service teachers' responses. Almost all of the pre-service teachers' responses to implementing a multicultural curriculum were enacted through a tourist approach. Respondents thought that using their potential future students and families as speakers/presenters and having world fairs or other isolated events meant an inclusive classroom and curriculum. While this approach was still present in the post-responses, pre-service teachers were beginning to develop a more sophisticated understanding of inequality and how it impacts student experience. It is important to help pre-service teachers move beyond a tourist approach where multicultural education is only a peripheral experience and move to holistically inclusive, diverse, and social justice oriented curriculum. Teachers and students need to take stronger action than just taking "trips" and visiting issues to acting as if multiculturalism is a critical part of life.

It is important to help pre-service teachers acquire the disposition that developing multicultural competency is a lifelong journey. One never "arrives" at multicultural competence. There is always additional work to be done. Developing a critical sense of self is essential to this work. Such awareness allows pre-service teachers to constantly evaluate and change the way they impact situations, as suggested in survey responses. Respondents realized they would have to do additional research. One goal of teaching multicultural competencies is to help pre-service teachers realize it is not just book knowledge (research) but also critical reflection and social/behavioral considerations that affect their growth.

This raises another important consideration: the themes used in the analysis were not designed to be mutually exclusive. Since multicultural competency is a lifelong journey, pre-service teachers' responses could contain elements of several categories. Colorblind understandings and deficit-based perceptions often were present in their definition of multicultural competencies, whereas cultural tourism was most often part of their implementation responses. The discussion of oppressed groups as the "other" was present throughout both the meaning and implementation responses. In many cases, pre-service teachers have colorblind understandings, deficit-based perceptions, and cultural tourism in the same responses, even when shifting toward pluralistic and social justice-based understandings.

Respondents to the survey demonstrated considerable growth in the area of multicultural competencies from the time they entered the program until they completed either EDR 514 or EDU 130. This is important because it shows the impact of college curriculum on pre-service teachers' perceptions and understanding of multicultural competencies. Pre-service teachers are able to reflect on their own experience and realize there is a need to shift their thinking about how to structure their own classroom in order to create a more democratic and just environment for future students.

One challenge present in the question framework was that the second question asked what the respondents would do to demonstrate their multicultural competencies in a classroom. The question framework could have impacted the frequency of responses that discussed activities and events. The goal for the program is to help pre-service teachers understand that multicultural competencies should undergird their curriculum and instructional practice. The response still provided information regarding woven multicultural education would be in their curriculum as future teachers.

10.6 CONCLUSION

While some have argued that schools are equalizers, providing level playing grounds for all students, critics contend that schools advantage certain groups based on their race/ethnicity, linguistic, gender, ability, or socioeconomic status (Collins, 2010; Halpern, 2011; Coleman, 1988; Murrell,

2006; Olneck, 2000; Banks and Banks 1989). The latter claim suggests that students of racial and ethnic minorities are disadvantaged by, disenfranchised from, and disempowered in schools. All P–12 students deserve to be in supportive and intellectually stimulating environments where they are encouraged to reach their fullest potential by teachers and peers. For these spaces to be nurtured, teachers need to do more than appreciate or accept diversity. They must continually develop their multicultural competencies and the multicultural competencies of their students.

This study reveals that the surveyed pre-service teachers have minimal multicultural competence. They often teach like they were taught, from monocultural experiences, and have avoided careful examination of uncomfortable topics such as classism, sexism, racism, and privilege. For pre-service teachers to successfully teach their diverse students, they must be reflective practitioners bent toward social justice. Teacher preparation programs have not only the opportunity but also the responsibility to help pre-service teachers develop multicultural competencies. In the process, teacher preparation programs can change the experiences and outcomes for racially and ethnically diverse students in P–12 schools.

KEYWORDS

- **Assimilation mindset**
- **Multicultural competency**
- **P–12 education**
- **Pre-service teachers**

REFERENCES

Ackland, R.; Schnackenberg, H.; Simard, D.; and Still, G.; *TEAC Inq. Brief.* **2013**. SUNY Plattsburgh.

Ambe, E. B.; Fostering multicultural appreciation in pre-service teachers through multicultural curricular transformation. *Teach. Teach. Educ.* **2006,** *22(6),* 690–699.

Anzaldúa, G.; Borderlands La Frontera. San Francisco, CA: Aunt Lute Books; **2007**.

Au, K. H-P.; and Mason, J. M.; Social organizational factors in learning to read: The balance of rights hypothesis. *Read. Res. Quart.* **1981,** *17(1),* 115–152.

Baltodano, M.; The accreditation of schools of education and the appropriation of diversity. *Cult. Stud. Critical Methodol.* **2006,** *6(1),* 123–142.

Banks, A.; and Banks, C. M.; Multicultural Education: Issues and Perspectives. Boston: Allyn and Bacon; **1989**.

Banks, J. A.; Educating Citizens in a Multicultural Society. 2nd Ed. New York: Teachers College Press; **2007**.

Banks, J. A.; An Introduction to Multicultural Education. 4th Ed. New York: Pearson; **2008**.

Benton-Borghi, B. H.; and Chang, Y. M.; Critical examination of candidates' diversity competence: rigorous and systematic assessment of candidates' efficacy to teacher diverse student populations. *The Teach. Educ.* **2012,** *47(1),* 29–44.

Bonilla-Silva, E.; Racism Without Racists: Color-Blind Racism & Racial Inequality in Contemporary America. 3rd Ed. Lanham, MD: Rowan & Littlefield Publishers; **2010**.

Cazden, C.; Classroom Discourse: The Language of Teaching and Learning. 2nd Ed. Portsmouth, NH: Heinemann; **2001**.

Cochran-Smith, M. (1995). Color blindness and basket making are not the answers: Confronting

the dilemmas of race, culture, and language diversity in teacher education. *American*

Educational Research Journal, 32(3), 493-522.

Coleman, J.; Social capital in the creation of human capital. *Am. J. Sociol.* **1988,** *94(Supplement),* S95-210.

Collins, P. H.; Another Kind of Public Education. Boston, MA: Beacon Press; **2010**.

Connell, R. W.; Masculinities. Berkley, CA: University of California Press; **1995**.

Darling-Hammond, L.; Chung, R.; and Frelow, F.; Variation in teacher preparation: how well do different pathways prepare teachers to teach?" *J. Teach. Educ.* **2002,** *53(4),* 286–302.

Davis, D. E.; Preparing White student teachers through a critical consultative interaction model. *Int. J. Prog. Educ.* **2009,** *5(2),* 23–43.

Delano-Oriaran, O.; Preparing pre-service teachers for culturally diverse classrooms: a conceptual model. *J. Educ. Teach.: Int. Res. Pedagogy.* **2012,** *38(1),* 103–106.

Delgado, R.; and Stefancic, J.; Critical Race Theory: An Introduction. New York: New York University Press; **2001**.

Delpit, L.; Lessons from teachers. *J. Teach. Educ.* **2006,** *57(3),* 220–231.

Evans-Winters, V. E.; and Hoff, P. T.; The aesthetics of white racism in pre-service teacher education: a critical race theory perspective. *Race Ethni. Educ.* **2011,** *14(4),* 461–479.

Feistritzer, C. E.; Griffin, S.; and Linnajarvi, A.; Profile of Teachers in the US 2011. Washington, DC: National Center for Education Information. Retrieved from http://www.ncei.com/Profile_Teachers_US_2011.pdf, **2011**.

Gomez, M. L.; Prospective teachers' perspective on teaching diverse children: a review with implications for teacher education practice. *J. Negro Educ.* **1993,** *62(4),* 459–472.

Gordon, J.; Inadvertent complicity: colorblindness in teacher education. *Educ. Stud.* **2005a,** *38(2),* 135–153.

Gordon, J.; White on White: researcher reflexivity and the logics of privilege in while schools undertaking reform. *The Urban Rev.* **2005b,** *37(4),* 279–302.

Gudykunst, W. B. and Kim, Y. Y. (1984) *Communicating with Strangers. An Approach to Intercultural Communication.*Random House, New York.

Halpern, D.; Social Capital. Malden, MA: Polity Press; **2011**.

He, Y.; and Cooper, J. E.; The ABCs for pre-service teacher cultural competency development. *Teach. Educ.* **2009,** *20(3),* 305–322.

Irvine, J. J.; Educating Teachers for Diversity: Seeing with a Cultural Eye. New York: Teachers College Press; **2003**.

Karafantis, D. M.; Pierre-Louis, J.; and Lewandowski, G. W. Jr.; A comparison of multicultural and colorblind perspectives on the intergroup attitudes of college students. *J. Hum. Behav. Soc. Environ.* **2010,** *20(5),* 688–710.

Kozol, J.; Savage Inequalities: Children in America's Schools. New York: Crown; **1991**.

Ladson-Billings, G.; Toward a theory of culturally relevant pedagogy. *Am. Educ. Res. J.* **1995,** *32(3),* 465–491.

Ladson-Billings, G.; and Tate, W. F.; IV. Toward a critical race theory of education. *Teach. Coll. Rec.* **1995,** *97(1),* 47–68.

Lobb, P. M.; Making multicultural education personal. *Multicult. Perspect.* **2012,** *14(4),* 229–233.

McAllister, G., & Irvine, J. J. (2000). Cross cultural competency and multicultural teacher education. *Review of Educational Research, 70*(1), 3-24.

McIntosh, P.; White Privilege: Unpacking the Invisible Knapsack. Retrieved from http://www.isr.umich.edu/home/diversity/resources/white-privilege.pdf, **1989.**

Milner, H. R.; Race, culture, and research positionality: working through dangers seen, unseen, and unforeseen. *Educ. Res.* **2007,** *36(7),* 388–400.

Mohatt, G.; and Erickson, F.; Cultural differences in teaching styles in on Odawa school: a sociolinguistic approach. In: Trueba, H.; Guthrie, G.; and Au, K.; Eds. Culture and the Bilingual Classroom: Studies in Classroom Ethnography. Rowley, MA: Newbury House; **1981,** 105–119.

Murrell, P.; Toward social justice in Urban education: a model of collaborative cultural inquiry in Urban schools. *Equit. Excell. Educ.* **2006,** *39,* 81–90.

Nieto, S.; Affirmation, solidarity and critique: moving beyond tolerance in education. In: Beyond Heroes and Holidays. Oakland, CA: AK Press; **2007**.

Olneck, M.; Can multicultural education change what counts as cultural capital. *Am. Educ. Res. J.* **2000,** *37,* 317–348.

Sogunro, O. A.; Toward multiculturalism: Implications of multicultural education for schools. *Multicultural Perspectives,* **2001,** *3(3),* 19–34.

Sleeter, C.; An invitation to support diverse students through teacher education. *J. Teach. Educ.* **2008,** *59(3),* 212–219.

Thompson, A.; Not the color purple: Black feminist lessons for educational caring. *Harvard Educ. Rev.* **1998,** *68(4),* 522–554.

Thompson, A.; Off White: Reading on race, power, and society. *Educ. Stud.* **1999,** *30(2),* 141–160.

US Census Bureau. US Census Bureau Projections Show a Slower Growing, Older, More Diverse Nation a Half Century from Now. Retrieved from http://www.census.gov/newsroom/releases/archives/population/cb12-243.html, **2012**.

INDEX

For Product Safety Concerns and Information please contact our EU
representative GPSR@taylorandfrancis.com
Taylor & Francis Verlag GmbH, Kaufingerstraße 24, 80331 München, Germany

www.ingramcontent.com/pod-product-compliance
Ingram Content Group UK Ltd.
Pitfield, Milton Keynes, MK11 3LW, UK
UKHW021959180926
475701UK00003B/105

* 9 7 8 1 7 7 4 6 3 2 8 5 7 *